Time and Causality Across the Sciences

This book, geared toward academic researchers and graduate students from multiple fields, brings together research on all facets of how time and causality relate across the sciences. Time is fundamental to how we perceive and reason about causes. It lets us immediately rule out the sound of a car crash as its cause, and learn which foods lead to illness. That a cause happens before its effect has been a core, and often unquestioned, part of how we describe causality. However, research across disciplines shows that the relationship is much more complex than that, with examples from quantum mechanics testing most existing theories. This book explores what that means for both the metaphysics and epistemology of causes – what they are and how we can find them. Across psychology, biology, physics, and the social sciences, common themes emerge, suggesting that time plays a critical role in our understanding, even when it is not explicitly mentioned. The increasing availability of large time series datasets allows us to ask new questions about causality, necessitating new methods for modeling dynamic systems and incorporating mechanistic information into causal models.

SAMANTHA KLEINBERG is an associate professor of computer science at Stevens Institute of Technology, New Jersey. She received her PhD in computer science from New York University, and previously held an NSF/CRA Computing Innovation Fellowship at Columbia University. She is the recipient of NSF CAREER and JSMF Complex Systems Scholar Awards and is a Kavli Fellow of the National Academy of Sciences. She is the author of *Causality, Probability and Time* (Cambridge University Press, 2012) and *Why: A Guide to Finding and Using Causes* (2015).

Time and Causality Across the Sciences

Edited by

SAMANTHA KLEINBERG
Stevens Institute of Technology

CAMBRIDGE
UNIVERSITY PRESS

University Printing House, Cambridge CB2 8BS, United Kingdom

One Liberty Plaza, 20th Floor, New York, NY 10006, USA

477 Williamstown Road, Port Melbourne, VIC 3207, Australia

314–321, 3rd Floor, Plot 3, Splendor Forum, Jasola District Centre,

New Delhi–110025, India

79 Anson Road, #06–04/06, Singapore 079906

Cambridge University Press is part of the University of Cambridge.

It furthers the University's mission by disseminating knowledge in the pursuit of
education, learning, and research at the highest international levelsof excellence.

www.cambridge.org
Information on this title: www.cambridge.org/9781108476676
DOI: 10.1017/9781108592703

First published 2019

Printed and bound in Great Britain by Clays Ltd, Elcograf S.p.A.

A catalogue record for this publication is available from the British Library.

Library of Congress Cataloging-in-Publication Data
Names: Kleinberg, Samantha, 1983–editor.
Title: Time and causality across the sciences / edited by Samantha Kleinberg,
Stevens Institute of Technology, New Jersey.
Description: Cambridge, United Kingdom : New York, NY, USA : University
Printing House, 2019. | Includes bibliographical references and index.
Identifiers: LCCN 2019009294 | ISBN 9781108476676 (hardback)
Subjects: LCSH: Science–Methodology. | Causation. | Time perception.
Classification: LCC Q175.32.C38 T56 2019 | DDC 501–dc23
LC record available at https://lccn.loc.gov/2019009294

ISBN 978-1-108-47667-6 Hardback

Contents

Contributors

Yin Chung Au
National Cheng Kung University

Inge de Bal
Ghent University

Neil R. Bramley
New York University

Phil Dowe
Australian National University

Tobias Gerstenberg
Massachusetts Institute of Technology

Victor Gijsbers
Leiden University

Jenann Ismael
Columbia University

David Jensen
University of Massachusetts Amherst

Samantha Kleinberg
Stevens Institute of Technology

David A. Lagnado
University College London

Bert Leuridan
University of Antwerp, Centre for Philosophical Psychology

Jonathan Livengood
University of Illinois at Urbana-Champaign

Thomas Lodewyckx
University of Antwerp, Centre for Philosophical Psychology and Ghent University, Centre for Logic and Philosophy of Science

Ralf Mayrhofer
University of Göttingen

Erik Weber
Ghent University

Naftali Weinberger
University of Pittsburgh, Center for Philosophy of Science

Karen R. Zwier
Iowa State University

1

An Introduction to Time and Causality

Samantha Kleinberg

1.1 Why Time and Causality?

Time is fundamental to how we perceive and reason about causes. It lets us immediately rule out the sound of a car crash as its cause, and learn through experience that alcohol may lead to hangovers. Our first instinct when asking why something happened is to look backward, aiming to identify an earlier event that may have been responsible. If you develop food poisoning and want to know how you became ill, you will likely start by enumerating all the things you recently ate. It probably seems obvious to you that you should only consider foods you consumed before you got sick and ignore those you ate afterward. At the same time, you would not bother trying to remember what foods you ate a week ago, last month, or last year. A cause needs to be earlier – but not too early. There is a range of times where we believe a food can plausibly lead to illness. If we want to explain a particular stock market crash, we again look backward to what happened leading up to the crash. The important events must still be earlier, but now we might consider a different window of time than for food poisoning, based on how we believe factors may influence the market and whether we believe this process is faster or slower than that for food-based illness.

Our reliance on time is not limited to explanation of specific events (token causality), but is central to how we learn about general relationships, such as what causes heart disease (type-level causality). To find causes of disease, we often examine what is different about people who become ill and those who do not. Once again, if people with heart disease tended to take up smoking after diagnosis, we would not consider this a cause of the illness, as we believe that only factors that happened before the onset of the disease could have caused it.

The order of events is one of the key features we use to separate potential causes from the large set of things we measure or observe. We are able to

1

split items into those that are potentially relevant and those that are not, solely based on when they happen. Of course, not every earlier event is responsible for every later event, but this belief about the importance of time provides a key heuristic for locating potential causes.

This intertwined relationship between time and causality may seem so obvious that it does not warrant examination, but it is not a given that time is such an important clue toward causality. There are many other factors we could use to weed out things that are not causes. We could consider, for example, which events have the capacity to cause others, which are nearby in space, or which are often responsible for a particular effect. For instance, when one billiard ball hits another we may suggest that the moving ball has momentum that it transfers to the second one or that the ball closest to the second is the one that caused its motion. While this is true, we immediately know, that even without any understanding of physics, the ball that was moving first caused the second ball to move. It is obvious to us what happened based solely on the order of events. We could similarly determine which foods are likely to be causes of food poisoning by identifying those that are most frequently responsible for food poisoning. This requires much more background knowledge and thought, though, than an initial filtering by time does.

Even though time is one of the key clues for humans trying to identify causal relationships, and even though it has been part of definitions of causality dating back centuries, it can also be misleading. It is such a strong clue to causality that we may believe one event caused another just because it happened earlier. This is known as the *post hoc ergo propter hoc* (after therefore because) logical fallacy. When we push a button at a crosswalk and the light immediately changes, we may think the timing is too much of a coincidence for the button to not be the cause of the light changing (Miele, 2016) or we may receive a flu shot and believe that a case of the flu shortly after was caused by the shot. The downside of time having such a strong influence on our understanding of causality is that this information may drown out other conflicting signals, such as knowledge that the flu shot contains an inactive form of the virus.

Psychologists have done experiments designed to create such a conflict between our temporal perception and our knowledge of how things must work. In cases where participants had full knowledge of a system and could predict how long it would take to operate, even though the real cause was slow, children mainly relied on temporal perception and claimed that the event happening immediately before the effect was its cause. Adults, on the other hand, were able to reason that such an event did not have time to produce the effect (Schlottmann, 1999). Yet in cases where there is not a fully observed mechanism, adults too find it less plausible that a cause is responsible for an effect as the time delay between them increases (Shanks, Pearson, and Dickinson, 1989). These errors may persist even when participants are told there may be delays and that the order in which events are observed may not be the same as the order in which they happen (Lagnado and Sloman, 2006).

In addition to erroneous perceptions of causality by humans, time may also lead to incorrect causal inferences by algorithms. Since the mid-1900s US gross domestic product (GDP) has grown exponentially. Many other things have grown at similar rates, such as smartphone usage and the number of Starbucks stores. Simply applying standard algorithms used to identify causal relationships to this raw data would likely result in finding any pair of these variables to be causally related. There is no relationship, though, except that these time series are all nonstationary. This means that key factors such as their mean or variance are changing over time. Because this trend overwhelms the smaller day-to-day or year-to-year changes, we can find spurious connections such as smartphone usage causing Starbucks to open more stores.

Nonstationarity is one of the fundamental challenges for inference from time series. In addition to finding spurious causal relationships, it can also make it more difficult to find true ones. For instance, patients change over time in the hospital as they recover, and interventions to treat illness may lead to changes in causal relationships such as how a medication affects physiology. Similarly, macroeconomic variables are influenced by policy changes, which may change behavior, and the locations being studied may change as country boundaries are not static. Even policies affecting human behavior, such as laws requiring individuals to wear seatbelts, may lead to fundamental differences in how people drive or what risks they take. In contrast to the case of nonstationarity leading to spurious inferences, here the causal relationships governing how these systems behave actually are changing over time. This leads to a moving target for inference, and many methods can only find the relationships that are stable across time, leading to missed inferences. Yet, finding specifically which relationships are changing – and how – is often more important than identifying those that are constant.

Even if a time series is stationary, there are other challenges for causal inference (finding causal relationships from data). A major problem is when the timing associated with data is at the wrong level of granularity, making events seem simultaneous when they are not. If causal influence is happening across time, and this order of events is important for finding causes, then inference becomes even more challenging when we cannot determine which event happened first. For example, if we measure GDP and inflation once a year, then if influence between these variables happens faster than yearly, we will have no idea which change came first and will only be able to find a correlation between the two time series.

In more challenging cases, different factors may be measured with different delays, leading to the events appearing out of order. Surprisingly, this situation is not uncommon. Continuous glucose monitors provide regular measurements of glucose for use in managing diabetes. These devices record glucose values every 5 minutes, but they measure glucose in fluid between the cells rather than in the blood stream. As a result, the measurements are delayed relative to blood glucose, on the order of 5–15 minutes (Bequette, 2010). Further the signals are not just shifted in time but rather the signal may be distorted in a way that depends on the current glucose value and its trajectory (Cobelli et al., 2016). Thus, it is possible that we may observe what seems like a meal and injection of insulin happening before a rise in glucose, solely because one variable (glucose) is measured with a delay that is not present for the others. This could lead to counterintuitive findings like insulin causing an increase in glucose if the delay is not taken into account or if too fine a timescale is used. Data that is changing, unreliable, or where the order of events cannot be trusted poses many challenges for learning causal models.

Despite these issues, a cause happening before its effect has been a core, and often unquestioned, part of how we describe causality, both in everyday life and in scientific research. However, understanding the relationship between time and causality is now an increasingly significant focus of research across many disciplines. This shift is driven in part by the availability of new datasets and measurement methods. Large observational datasets such as from medical records and social media enable us to ask new types of questions on causality. They also bring new challenges in identifying causal relationships, as we face missing variables, noise, and causal relationships that change over time. These challenges are common to research in many fields, and their solutions may in turn provide insight into core philosophical problems. For example, we can now examine risk factors for disease using electronic medical records going back decades, from much larger populations than were previously possible. Yet the meaning of variables changes over time (what was considered light smoking in the 1970s may not mean the same thing today), as do populations and their behaviors. Similarly, more sensitive measurement techniques may now let us determine whether events thought to be simultaneous actually have a time order. These in turn may shed light on questions such as whether a cause coming before an effect is necessary or just the most likely order of events. Given these many opportunities and challenges, it is an ideal time to examine the relationship between time and causality in depth.

1.2 Interdisciplinary Perspectives on Time and Causality

This book brings together research on all facets of how time and causality relate across the sciences. Through these interdisciplinary perspectives, we explore the role of time in philosophical theories of causality, how time affects our perception and judgment of causality, and new methods for inferring causes from temporal data. This section provides historical context for the three main segments of the book (philosophy, sciences, and models and algorithms) and brief overviews of the chapters, with a focus on discussing how they illuminate different aspects of the same core problems.

1.2.1 Philosophy

In the eighteenth century, David Hume (1748, Section VII, Part II; emphasis original) defined a cause as *"an object, followed by another, and where all the objects similar to the first are followed by objects similar to the second. Or, in other words, where, if the first object had not been, the second never had existed."* More than two hundred years later, these definitions are the foundation of two main philosophical approaches to defining causality, as well as the computational methods for finding it that will be discussed later in this book. The first sentence says causality is about regularities, where the effect routinely follows the cause. The second elaboration provides the basis for the counterfactual approach to causality, where we examine what would have happened had the cause not occurred (see Lewis, 1973a). That is, rather than a repeated sequence of events, here the effect must make a difference to whether or not the cause happens. While there are many differences between the two definitions, a key commonality is that both assume that there is a first event (the cause) and a second (the effect). This requirement that a cause be earlier than its effect is known as temporal priority. Hume also stipulated that causally related events must be contiguous (nearby in space and time) and that the cause is required for the effect to occur (necessary connection). In this book we focus primarily on the temporal aspect of this definition.

In practice, effects may not always follow their causes. While smoking can cause lung cancer, not everyone who smokes develops cancer, and some non-smokers also develop cancer. One source of uncertainty may come from our observations (perhaps if we knew enough about what caused cancer, we could specify the cause such that every time it was present cancer resulted), while the other is that some causes are at their core non-deterministic. That is, in some cases even with complete knowledge about the scenario, we cannot be sure whether the cause will produce the effect. As a result, research stemming from the regularity theory has aimed to specify these conditions more fully to distinguish causal regularities from correlations (Mackie, 1974/1980), or incorporate probability into the definitions of causality.

In the 1950s, Hans Reichenbach (1956) attempted to define causality in terms of probability and, uniquely, to learn the direction of time from the direction of causality. That is, in his theory a common cause that fully explains its effects must be earlier than the effects, and by finding this explanatory event we can determine which way time flows – rather than using the direction of time to determine which events may be potential common causes. Reichenbach's theory is unique in this respect, using the asymmetry of probabilistic relationships between cause and effect to identify causes without knowing which is earlier. However, without the use of temporal information, one must make other, stronger, assumptions. Thus, Reichenbach proposed definitions for a common cause of two effects based on their probabilities, with the stipulation that the effects are simultaneous. If there is no earlier common cause, this ensures we will not find one event as a cause of the other. For example, if two lamps go out at the same time, we will not erroneously find one lamp's state causing the other's. Yet for time series data that is nonstationary we may find two series with similar trends where neither is a cause of the other and yet there is no common cause (violating Reichenbach's common cause principle).

While Reichenbach's theory influenced many later probabilistic theories, work on probabilistic causality returned to the use of temporal priority as a distinguishing feature of causality. Patrick Suppes' influential work on distinguishing mere events that raise the probability of their effects from genuine causes assumed we know the ordering of events and then defined ways in which earlier events may make later ones spurious (Suppes, 1970). Similarly, while Ellery Eells (1991) developed measures for probabilistic causal relevance that consider factors happening temporally in between the cause and effect, he again only considers causes that are strictly earlier than their effects. Note, though, that while Hume's initial goal was to uncover both the metaphysics (what causes are) and epistemology (how we can learn about causes), these probabilistic theories are primarily focused on epistemology. This is an important distinction when examining the relationship of time and causality, since it is possible that temporal priority is an effective clue toward identifying causes, even if it is unnecessary for something to actually be a cause.

Thus, while time is a core part of how people think about causes, is it essential for a definition of causality? According to Hume it is a part of causality itself, but it is also possible that while temporal order is a useful heuristic for people to identify causes, it may not be a fundamental part of what it means to be a cause. As discussed later in this book, physics has provided seeming counterexamples to this stipulation with cases where the future may potentially affect the past. For example, Price (2012b) argues that some quantum theories must allow retrocausality. Reichenbach argued that we can instead identify the order of time from the direction of causality. While less extreme than causality flowing backward in time, there has been compelling new evidence of simultaneous causation in psychology and neuroscience (Vernon, 2015).

The first segment of the book addresses the philosophy of time and causality, exploring what this new evidence means for both the metaphysics and epistemology of causes – what they are and how we can find them. One of the fundamental questions about the relationship between causality and time is whether causes must precede their effects. Even without backwards causality (causes affecting events that happen earlier than they do), we may ask whether all causes are indeed strictly earlier than their effects and whether this temporal precedence is essential to a theory of causality. In Chapter 2, Thomas Lodewyck and Bert Leuridan begin with a detailed examination of the role of time in all major theories of causality, providing a road map for all later chapters. They argue for an account of causation that is independent of time. Notably, the core of this argument is that existing philosophical theories that aim to identify causal direction from temporal direction or vice versa cannot adequately handle scientific phenomena. Phil Dowe, in Chapter 3, focuses on the relationship between time and metaphysics and the puzzle of how we can answer what it means for one event to cause another in so many different ways. Dowe shows that many answers are lacking and discusses what is needed to fully answer this question. While many arguments exist for why temporal (and even spatial) ordering is necessary for understanding causality, what about the reverse: must we have an understanding of causality to have a concept of space and time? That is, can there be worlds that proceed temporally but without causation? In Chapter 4, Victor Gijsbers argues that causality and time are so interlinked that they cannot be analyzed separately, suggesting that a new approach to both causality and time is required. In Chapter 5, Jenann Ismael discusses how recent advances in physics, computer science, and decision theory have provided new insights into these old philosophical problems, and prompted new questions. In particular, she ties together the philosophical study of causality, evidence from physics, and computational methods for inferring causes. A common theme throughout the volume is the use of methods and findings from other fields to advance understanding across disciplines.

1.2.2 Sciences

Hume and others have provided the foundations for how we understand the metaphysics and epistemology of causality. Many scientific disciplines, though, are primarily concerned with the methodology used to find causes or support causal claims. That is, regardless of what a cause actually is, what are the tools we can use to identify such relationships? Even though it is not always explicit, causality is at the core of most scientific pursuits, which aim to learn how things work. For example, biologists have asked questions such as what happens during the cell cycle, what functions different genes have, how organisms have evolved, and so on. The answers to these questions may involve uncovering specific labels for processes (e.g., mitosis is one phase of the cell cycle, and is when the cell divides) or links between events (e.g., a particular gene is necessary for a specific eye color), rather than explicitly causal language. Yet the goal is still uncovering why and how things happen, which are both causal questions (Mayr, 1961).

All of these processes unfold over time, though this is often implicit in the explanations given in biology. While there may be regularities at work, the way causes are described in the biological sciences bears more similarities to mechanistic theories of causality. Rather than determining whether or not there exists a causal link between events, these approaches instead focus on identifying the processes by which the cause brings about the effect (Glennan, 1996a; Machamer et al., 2000). A mechanism can be thought of as a set of parts that interact to regularly produce a change. Time is implicit here, as the process must have some duration at the end of which the effect is present.

In biomedical sciences, such as epidemiology and public health, the distinction between causality and correlation not only leads to deeper understanding of phenomena, but is fundamental for making policies and choosing medical interventions. A chemical may be found in cigarettes (and thus associated with cancer) but may not be a carcinogen on its own, and so different determinations about its safety may be made. Even though causality is critical for decision-making here, it is much more challenging to establish than in biology since it is rarely possible to conduct controlled experiments, many factors interact to affect health, these relationships are rarely deterministic, and often they unfold over long time periods. As a result, there has been much effort toward clarifying how causal claims can be made. Austin Bradford Hill (1965) proposed a set of factors to consider when evaluating causal claims, though these have been frequently misrepresented as criteria for causality. One of the factors proposed is temporality (does the cause happen before the effect?), yet Hill notes that in epidemiology this feature may not be as straightforward as it seems. For example, individuals with chronic heart failure have reduced exercise tolerance. Yet, given that heart failure takes time to develop (and we rarely see it in its early stages), we may observe the impact on exercise before heart failure is actually apparent and diagnosable – leading to a mistaken temporal impression and the idea that reduced exercise is causing heart failure.

The second part of this volume delves into the relationship between time and causality in specific disciplines, including psychology, biology, and physics. Work in psychology has examined many facets of our understanding of causality, including when this understanding develops in children, whether we are able to perceive causality directly, and how causality and moral considerations interact in assigning blame. There has been a large amount of work on how we learn about causes, and in particular how being able to experiment on a system (intervene) affects causal learning, and how temporal cues are used. In Chapter 6, Neil Bramley and others propose a unification, showing that it is difficult to separate time and causality both conceptually and experimentally and discussing how both affect our ability to learn in complex situations. This research into what affects human judgments of causality may provide insight that can guide algorithms that discover causal relationships from data. In biology, time is a key factor in understanding the mechanisms by which processes occur, yet time is often implicit. Yin Chung Au, in Chapter 7, argues that even when it is not explicitly mentioned, time plays a critical role in the search for biological mechanisms. Using the specific example of apoptosis (programmed cell death) studies from the 1970s to the present, the chapter examines both how biological mechanisms are identified using temporal information and the role of time in diagrams used for communicating and explaining these mechanisms. Much like biology, work in physics illustrates how pervasive the assumption of a temporal order is. Inge de Bal and Erik Weber, in Chapter 8, provide new insights into how time can be used to identify causes in ways that may translate across disciplines. They investigate the relationship between mechanistic and correlational evidence in reasoning about physical causal claims. In particular, they show that mechanistic evidence can explain how we go from symmetric physical laws to asymmetric causal claims. While de Bal and Weber build on analogies from the social sciences, Jonathan Livengood and Karen R. Zwier show, in Chapter 9, that time is often not explicitly modeled in the social sciences. This chapter shows how this can lead to models where causes do not precede their effects and proposes criteria to both preserve the intended ordering and remain consistent with the data analyzed. Using examples from sociology, political science, and education research, they show that these criteria are non-trivial and that researchers must carefully consider the commitments of the models they develop even – or especially – when time is not explicit.

1.2.3 Models and Algorithms

While uncovering causality has been a core goal of many sciences and a topic of inquiry for centuries, it was only in the 1980s that the computational methods for modeling causal relationships and inferring them from data emerged. A major advance was the work of Judea Pearl (2000) and Peter Spirtes, Clark Glymour, and Richard Scheines (Spirtes et al., 2000) who showed that probabilistic causal relationships could be represented using graphs called Bayesian networks. They defined what makes a graphical model a causal one, and what we need to assume to draw causal conclusions. Further, they created algorithms for learning these models from data. While these methods are highly general, focused on identifying relationships between variables, (e.g., smoking causes lung cancer) they do not define the timing of these relationships (e.g., does smoking cause lung cancer in 10 years or 20 years?) or what the meaning of a variable is (e.g., what does it mean for smoking to be true – smoking for a month, a year, 10 years? What quantity must one smoke?). While the methods described could use temporal data to orient edges in the graph from causes to effects, they had no way of learning the timing for each relationship, and further had to make many assumptions that are not widely true in time series data, such as that relationships are stable over time. More recent work has developed dynamic Bayesian networks, which extend these methods to include a notion of time, using a sequence of Bayesian networks connected across time to show how factors at one time influence those at a later time (Murphy, 2002).

Nevertheless, there has been significantly less attention paid to the much more challenging temporal case. The challenges are both conceptual and practical, including how to identify causes at different levels of temporal granularity and developing efficient methods given the increased complexity of inference in the temporal case. Despite limited computational work, understanding causal relationships over time is of significant practical importance. For example, to develop effective economic policies we must be able to compare both the cost of various strategies as well as how long they will take to achieve their goals. Further, many domains produce time series data. As a result, one of the earliest formulations of how to identify potential causal relationships in time series came from the economist Clive Granger (1980). Granger causality provides a way of identifying relationships between time series (roughly using how much information the history of one time series provides about another), but cannot be guaranteed to find causes. That is, there is no set of conditions under which a Granger-cause can be guaranteed to be a genuine cause. There are many cases where such causes may be spurious, such as when two time series are analyzed, but they are caused by a third variable. Despite these challenges, because of the simplicity of the approach it has been applied to a number of applications both in economics as well as for analysis of neurological time series including fMRI (Bressler and Seth, 2011).

While these models and algorithms are discussed throughout the earlier chapters of the book, the last two chapters focus on core challenges and open problems. First, Naftali Weinberger, in Chapter 10, examines models that are explicitly designed for temporal data, discussing the challenges of nonstationarity and what happens when variables are by definition linked. For instance, a billiard ball's velocity provides a lot of information about when and if it will end up in the pocket. While intuitively we may understand this relationship causally, it can provide challenges for modeling approaches such as those based on interventions. This chapter argues for the importance of causal modeling of conceptually related variables, and in particular examines the case of causal models that include variables and their time derivatives. Lastly, while Bayesian networks have been a dominant paradigm for causal inference, recent work has shown that they may not be sufficient to express the complex relationships found in temporal data. In Chapter 11, David Jensen examines the gap between what can be expressed in these models versus what explanations are needed by practicing scientists. As de Bal and Weber discuss in the context of physical claims, Jensen argues there is a need for mechanisms in causal modeling. This chapter surveys recent work and future directions toward richer descriptions of mechanisms.

1.3 What's Next?

This book aims to provide an introduction to the depth and breadth of work on time and causality. Throughout this interdisciplinary volume, some common themes emerge. As philosophers take inspiration from findings in physics and information theory, so too have computer scientists used philosophical work such as on mechanisms to further computational advances. Despite the increasing interest in time and causality, many problems remain unsolved though. Here I briefly highlight some of the core challenges and open problems for this area of work.

There is a clear need for collaboration beyond disciplines and even perhaps a feedback loop between philosophy, computation, and specific scientific domains. In practice, research often remains stuck within disciplinary conferences and journals. Yet, understanding what data is actually available can motivate a new understanding of the epistemology of causality. Similarly, insight into the metaphysics of causality can provide inspiration for algorithms to discover it from data. While metaphysical problems are different from epistemic and methodological ones, understanding of metaphysics can provide inspiration for new methods, and similarly the result of these new methods may provide counterexamples to metaphysical claims.

Many of the computational problems that have been addressed for the static case (or when ignoring time), become vastly more challenging when time is involved and remain unsolved. For example, a core open problem is developing algorithms to learn causal structures from data with latent variables. If A causes B and C, and A is not measured, we might incorrectly find that B causes C or vice versa. While there have been some solutions proposed, none have been widely effective in temporal data. This is only one of many problems where solutions exist – but only when time is not involved. Yet, as shown throughout this book, even when time is not explicitly modeled, it implicitly exists, and failing to account for it can lead to erroneous results. On the other hand, as we learn more about what is possible through better measurement techniques, theories may need to change, such as adapting to account for simultaneous causation. More broadly, we need a better understanding of the relationship between time and causality beyond just the order of events. Fundamental open questions remain, such as how to relate causal relationships at multiple timescales, what inferences can be made when variables are observed out of order, and what even constitutes a variable (or the "right" set of variables).

Lastly, one of the major gaps in the study of causality and time has been the integration of human understanding. While there has been a growing body of work in psychology on how time affects our ability to reason about and learn about causality, this work has not yet formed a feedback loop with the study of what causes are and how to find them. In particular, while there are many algorithms for finding causes from data, we do not know how useful these are for human decision-making. That is, is the most complete and correct causal model the one that is most useful to human decision-making? Without understanding who is using the output of these methods, we cannot adequately evaluate their utility. Instead, by linking work on how people use causal information to make decisions (whether in daily life or for creating policies) with work on extracting causal relationships from data, we can better support these decisions.

Acknowledgments

This volume is based on a selection of papers presented at Time and Causality in the Sciences, held at Stevens Institute of Technology in Hoboken, NJ in 2017. The conference brought together attendees from biology, computer science, philosophy, psychology, the social sciences, and statistics (among other fields), to focus on all aspects of the relationship between time and causality as it relates to scientific work. I want to thank all presenters and participants in the conference, who made the discussions so thought-provoking. Special thanks to Phyllis Illari, an entity whose activities were extremely organized and produced many phenomena in this book. The conference was supported in part by the National Science Foundation and Stevens Institute of Technology, and this research was also supported in part by the James S. McDonnell Foundation.

2

Causality and Time: An Introductory Typology

Bert Leuridan and Thomas Lodewyckx

2.1 Introduction

Causes precede their effects. Effects come after their causes. I strike a match; shortly after, it lights. John Jones has been smoking heavily during his adolescence; decades later, he develops lung cancer. An embryo inherits genes from both its parents; in the course of its subsequent ontogeny, it develops a range of phenotypic traits. It would not make sense to say that the lighting of the match was a cause of the striking, or that Jones's lung cancer caused him to smoke, or that an organism's phenotypic traits causally determine its genome: *propter hoc, ergo post hoc*. At least, that's what common sense tells us.

The common sense idea that causes must precede their effects is fairly widespread in the philosophical literature. This is illustrated in the following quotes:

> Now even the most daring thinker would hardly recommend the consumption of omelettes as a way of inducing hens to *have laid* their eggs. Indeed, in this form of words, the absurdity of post-dating causes is flagrant. (Black, 1956, p. 58, original italics)

> It doesn't matter whether the effect coincides with the last of its conditions, or immediately follows it. It doesn't precede it; and when we are wondering which of two coexistent phenomena is the cause and which the effect, we rightly regard the question as answered if we can ascertain which of them came first. (Mill, 1843/1911, Book III, chapter V, §7)

Causality is a notoriously difficult notion. What precisely does it mean to say that *x* causes *y*? The quest for a definition has haunted philosophers and empirical scientists for centuries. It is related to other important issues such as how we can know or discover that *x* causes *y* (epistemological or methodological questions) and how we can put such causal knowledge to use, for instance to intervene in the course of events (practical questions).

In this chapter we will focus on the first, definitional or conceptual question: What precisely does it mean to say that x causes y? Our aims are two-fold. First, we want to offer an introductory overview of a number of philosophical definitions or theories of causality. This overview should in particular be accessible to non-philosophers. Second, we will explore how the relation between causality and time is portrayed in these definitions and theories.

Given the common sense idea that causes must precede their effects, and its adoption in the philosophical literature, it is natural to expect philosophical theories to define causality in terms of temporal priority (plus some other conditions). In Section 2.2 we will present some *time-first accounts of causation* which do exactly this (Hume, Mill). Some philosophers have done the reverse, however. In Section 2.3 we will discuss a *causality-first account of time*, i.e. a theory of time that is based on a prior notion of causality (Reichenbach). It goes without saying that this prior notion of causality must be defined without invoking time. In Section 2.4 we will go through some theories of causality that are likewise defined without invoking time, yet without the goal of serving as the ground for a theory of time (Mackie, Salmon, Dowe, Menzies and Price, Lewis, and Woodward). Time and causality are treated in a more or less independent fashion here. Hence, we call these *time-independent accounts of causation*. In Section 2.5 we will touch on two further issues concerning the relation between causality and time: instantaneous causation and backward causation. In Section 2.6 we briefly defend our preference for time-independent approaches over their time-first and causality-first alternatives.

2.2 Time-First Accounts of Causation

2.2.1 David Hume's Regularity View of Causation

The eighteenth-century philosopher David Hume has offered a very well-known and influential definition of causality.[1] To be more precise, he has offered (at least) two non-equivalent definitions. In this section we focus on one of these. It is best understood by means of a concrete example repeatedly used by Hume himself: the collision of two billiard balls.

---

[1] Hume first published his theory of causation, among other things, in his *Treatise of Human Nature* (1739/1978). In 1740 he published *An Abstract of 'A Treatise of Human Nature'* (1740/1978). In 1748 he published the *Enquiry Concerning Human Understanding* (1748) in which he repeated portions of the *Treatise* in a more popular fashion. (See Fieser, 2018, for an accessible overview of Hume's work.)

Here is a billiard-ball lying on the table, and another ball moving towards it with rapidity. They strike; and the ball, which was formerly at rest, now acquires a motion. This is as perfect an instance of the relation of cause and effect as any which we know, either by sensation or by reflection. Let us therefore examine it. (Hume, 1740/1978, p. 649, as quoted by Psillos, 2002, p. 20)

Hume was one of the champions of empiricism, the philosophical doctrine "that all knowledge is ultimately based on experience" (Alston, 1998, article summary). Hence our knowledge of causal relations must be based on sensory experience, either directly (via impressions) or indirectly (via memories which we have of past impressions). Now what impressions do we have when we watch the billiard balls moving? First, we see that the two balls touch one another (in space) and that there is no interval (in time) between the shock and the motion of the second ball. In Hume's words: "*Contiguity* in time and place is therefore a requisite circumstance to the operation of all causes" (Hume, 1740/1978, p. 649, as quoted by Psillos, 2002, p. 21; emphasis original). Second, we also see that the motion of the first ball comes before the motion of the second ball and not the other way around. "Priority in time, is therefore another requisite circumstance in every cause" (Hume, 1740/1978, p. 650, as quoted by Psillos, 2002, p. 21). Finally, we recall that in similar cases, like causes were followed by like effects. In other words, whenever one ball moved with rapidity and collided with a second ball, that second ball acquired a motion. "Here therefore is a *third* circumstance, *viz.*, that is a *constant conjunction* betwixt the cause and effect. Every object like the cause, produces always some object like the effect" (Hume, 1740/1978, p. 650, as quoted by Psillos, 2002, p. 21).

This is all there is to causation, Hume writes. No extra ingredient is needed in our definition of causality. In particular, no necessary connection between cause and effect must be presupposed. (Have you ever *seen* that an effect *must* follow its cause? No, we only *see* that it *does* follow and we *recall* that it has done so every time in the past.)

In view of its second clause, Hume's regularity view is rightly called a time-first account of causation. It defines causation in terms of, among other things, temporal priority.

2.2.2 John Stuart Mill: Sophisticating Hume's Regularity View

The work of David Hume, and his empiricist stance in particular, has been very dominant in twentieth-century debates on causality and related issues. For a long time, there was a quasi-consensus among philosophers of science that no necessary connections should be posited in nature. His definition of causality, however, stood in need of refinement. Many cases of cause and effect do not satisfy Hume's original account. In particular, its third clause is too simplistic. My striking a match causes it to light, but it is not the case that *every time* a match is struck, it lights. Not *everyone* who smokes heavily during adolescence develops lung cancer later in life. And the relations between genes and phenotypes can rarely be modelled as simple cases of constant conjunction.

The nineteenth-century empiricist philosopher John Stuart Mill offered the following diagnosis for this problem. What we call the cause is seldom sufficient for producing the effect. The right background conditions need to be in place. Some conditions must be present (Mill calls these positive conditions), others must be absent (negative conditions). In the case of the match, oxygen must be present and the match may not be humid, among other things. In *A System of Logic*, originally published in 1843, he writes the following:

> It is seldom, if ever, between a consequent and a single antecedent that this invariable sequence [posited by Hume] subsists. It is usually between a consequent and the sum of several antecedents; the concurrence of all of them being requisite to produce, that is to be certain of *being followed by*, the consequent. (Mill, 1843/1911, p. 214, our emphasis)

Like Hume, Mill is very clear that causes must precede their effects. He distinguishes two relations between phenomena in nature, that of simultaneity and that of succession:

> Every phenomenon is related, in an uniform manner, to some phenomena that coexist with it, and to some that have preceded and will follow it. (Mill, 1843/1911, Book III, chapter V, §1)

The succession of phenomena (and only the succession of phenomena) is governed by the Law of Causation. Hence Mill's sophistication of Hume's regularity view is a time-first account of causation as well.

2.3 A Causality-First Account of Time: Hans Reichenbach's Probabilistic Causality and the Direction of Time

Causality-first accounts of time explicate time in terms of causality, not vice versa. The best known example is Reichenbach's. Another example, which we will not discuss, is Tooley's (1997, ch. 9). Later, Huw Price (1992, 1996) would come close to one, yet we discuss his work with Menzies under the heading of time-independent accounts of causality (see Section 2.4.3).

As a leading philosopher of science and empiricist in his time, Hans Reichenbach (1891–1953) has had an enduring impact on the philosophical treatment of causality. The ideas he developed in *The Direction of Time* (1956/1971) stood at the cradle of several modern-day accounts of causality.

As the title of his work clearly shows, Reichenbach was, above all, interested in the direction of time, a notion that was very much at the center of discussions in both philosophy and physics. It was clear from philosophical treatments that there exists a definite psychological separation of time into past and future. We can remember the past, but obviously not the future. From a physical viewpoint, however, this asymmetry does not seem to hold. At the level of the micro-processes, the laws of classical mechanics were generally thought to be time reversible, while macro-processes are irreversible. Reichenbach's most influential line of reasoning concerning the arrow of time intended to produce such a direction from causal directions inferred from macro-statistics, making him a pioneer of probabilistic causality. (See Illari and Russo, 2014, pp. 76–82, for a discussion of probabilistic causality.)

Here is a clear statement of his causality-first endeavour:

> How can we find a suitable explication of time? It is clear that it can be sought only by a study of the relationships of causality. The close connection between time order and causal processes will be evident from the preceding discussion. It will be the aim of the present investigation to show that this is more than a mere coincidence, that time order is reducible to causal order. Causal connection is a relation between physical events and can be formulated in objective terms. If we define time order in terms of causal connection, we have shown which specific features of physical reality are reflected in the structure of time, and we have given an explication of the vague concept of time order. (Reichenbach, 1956/1971, pp. 24–25)

Reichenbach (1956/1971, p. 25) traces this causality-first approach to the direction of time to the work of the German philosopher Gottfried Wilhelm Leibniz (1646–1716) and motivates it by referring to Einstein's criticism of simultaneity.

Reichenbach was well aware that time-first ideas of causality are very common and that this poses a challenge to his endeavour:

> In the experiences of everyday life we take it for granted that the cause-effect relation is directed. We are convinced that a later event cannot be the cause of an earlier event. But when we are asked how to distinguish the cause from the effect, we usually say that of two causally connected events, the cause is the one that precedes the other in time. That is, we define causal direction in terms of time direction. Such a procedure is not permissible if we wish to reduce time to

causality; and we therefore must look for ways of characterizing the cause–effect relation without reference to time direction. (Reichenbach, 1956/1971, p. 27)

In order to characterize causality without reference to time direction, Reichenbach opted for a probability interpretation of causality. He was the originator of the Principle of the Common Cause, which he formulated as follows: "*If an improbable coincidence has occurred, there must exist a common cause*" (Reichenbach, 1956/1971, p. 157, original italics); and as follows: "In order to explain the coincidence of *A* and *B*, which has a probability exceeding that of a chance coincidence, we assume that there exists a common cause *C*" (Reichenbach, 1956/1971, p. 159, original italics). The existence of the common cause is not absolutely certain, but it is probable in view of the coincidence.

The principle can be illustrated with the following example:

Suppose that lightning starts a brush fire, and that a strong wind blows and spreads the fire, which is thus turned into a major disaster. The coincidence of fire and wind has here a common effect, the burning over of a wide area. But when we ask why this coincidence occurred, we do not refer to the common effect, but look for a common cause. The thunderstorm that produced the lightning also produced the wind, and the improbable coincidence is thus explained. (Reichenbach, 1956/1971, p. 157)

This idea is spelled out more technically as follows. A conjunctive fork (in present-day terminology a common cause structure $A \leftarrow C \rightarrow B$) is defined by Reichenbach as an ordered triple of events ACB satisfying the following four relations:[2]

- $P(A \times B/C) = P(A/C) \times P(B/C)$
- $P(A \times B/{\sim}C) = P(A/{\sim}C) \times P(B/{\sim}C)$
- $P(A/C) > P(A/{\sim}C)$
- $P(B/C) > P(B/{\sim}C)$

A conjunctive fork ACB is closed at C, in Reichenbach's terminology. At the 'other side', it can be open or closed.[3] In theory, conjunctive forks may be open either to the future or the past. According to Reichenbach, however, it is a contingent fact about our world that we will only ever find open conjuctive forks that are future directed (see Dowe, 2000, p. 194). Reichenbach uses this idea in an attempt to define the direction of time.

---

[2] See (Reichenbach, 1956/1971, p. 159); we have changed the notation.
[3] Reichenbach defines this as follows: "If there is no event E on the other side of the fork which satisfies [the four relations], we say that the fork is open on that side" (1956/1971, p. 162).

DEFINITION: In a conjunctive fork ACB which is open on one side, C is
earlier than A and B. (Reichenbach, 1956/1971, p. 162)

He adds: "This is a definition of time direction in terms of macrostatistics"
(Reichenbach, 1956/1971, p. 162).[4] (It was correctly remarked by a reviewer,
however, that this sentence doesn't actually read like a definition, but rather as
a contingent if–then statement.)

Reichenbach was also among the first to suggest a type of causal linking.
This involved a mark, resulting from an intervention by means of an irreversible process. The mark is transmitted down causal chains and manifests
at each subsequent point. (Reichenbach, 1956/1971, §23; see also Illari and
Russo, 2014 pp. 112/138–139; and see Section 2.4.2 for a recent elaboration
of this idea.) In this way, there would be an asymmetry in the way causality
is imparted. In such a view, a process is causal only if it is marked at an
earlier point, with the mark necessarily being transmitted to later points in
the process.[5] A mark added to a point at the end of a chain will not show up
towards its beginning. Consider a racing pigeon, marked in the way we have
just described, travelling from Calais to London. At every stage of its journey
the pigeon will stay marked. However, if we only mark the pigeon in London
the mark will not be present in Calais or at any intermediate point before the
pigeon's arrival in London. Analogous to the psychological separation of time
into past and future, this implies a clear physical, and causal, asymmetry.

2.4 Time-Independent Accounts of Causality

Compelling as the common sense idea that causes must precede their effects
may be, many philosophers have refrained from building temporal priority
into their definitions of causality, even though their primary goal was not to
offer a causal account of time. In our opinion, there are three good reasons
for preferring time-independent accounts of causality over time-first accounts.
These reasons are repeatedly cited by time-independent theorists of causality.
The first reason is to reduce the conceptual cost of your account. If causality
can be defined without invoking the notion of temporal priority, why would

---

[4] We leave out much of the technical details of Reichenbach's exposition (see §19 of his book).
Reichenbach turns out to be confused about some of the relations between causality and
probability that are now well understood in the causal modelling literature. See Glymour and
Eberhardt (2016, §4.6) for a discussion of Reichenbach's mistakes; see also the standard works
in the causal modelling literature by Pearl (2000) and by Spirtes, Glymour and Scheines (2000).

[5] The use of the words 'earlier' and 'later' does not assume a preexisting time order. The time
order is guaranteed by the irreversibility of the marking process (Reichenbach, 1956/1971,
pp. 198–200).

you invoke it? Doing so would, strictly speaking, require that you also have an adequate theory of temporal priority, and hence of time.[6] The second reason is that it leaves open the possibility of using your account of causation to establish a causal account of time – even if that is not your own project. Third, it makes your account less vulnerable to the possibility of backward or instantaneous causation, where effects respectively precede or occur simultaneously with their causes (more on this below).

2.4.1 Mackie: Causes as INUS Conditions (Sophisticating Hume's Regularity View Even Further)

John Mackie (1974/1980) further elaborated on Mill's account to formulate an even more sophisticated version of Hume's regularity view, even though he was only interested in developing this account in an effort to properly criticize it. Mackie adopted the Millian view that effects can be brought about or *caused* by a cluster of factors or conditions (Psillos, 2002, pp. 87–92). Yet he added that an effect may result from a number of possible complex causes. Take, for example, a completely burned down house. There are a number of clusters of factors that can cause house fires. One cluster may represent some sort of electrical short circuit, combined with the presence of inflammable material and a lack of custodial supervision, while another cluster may include an arsonist, some matches and a drum of gasoline.

Each of these clusters could be said to be *sufficient* to cause the burned out building, but none of them are *necessary*. If the cluster including the short circuit obtains, the building will burn (the cluster is sufficient); yet the building may also burn in the absence of that cluster (the cluster is not necessary), as long as another sufficient cluster obtains. Within a cluster, as claimed by Mill, any one factor (e.g. the presence of flammable material, a lack of custodial supervision and so forth) is *insufficient* to singlehandedly cause the effect. Without the right set of other conditions in place, the presence of an oil drum does not lead to a burned down house. A cluster also should not contain any unnecessary factors (e.g. the specific color of the drum of gasoline), making all the factors present *non-redundant*.

The above example case serves to illustrate Mackie's theory of causation: a cause is taken to be "an *insufficient* but *non-redundant* part of an *unnecessary* but *sufficient* condition" (Mackie, 1974/1980, p. 62, original italics, underlines added). Causes are so termed as INUS-conditions.[7]

---

[6] Note that this advantage needs a qualification: if the added machinery needed to distinguish causes from effects is more 'expensive' than a theory of time, no conceptual cost is saved.

[7] According to Mackie, each single INUS-condition may be called a cause. This differs from Mill's formulation.

Even though Mackie presents a further sophistication of Hume's and Mill's regularity views of causation, his is not a time-first account. Temporal priority of causes to their effect is not part of his definition of causation. He even leaves open the possibility that causes chronologically follow their effects:

> But temporal priority is not conceptually necessary for causal priority. We can coherently consider the possibility of backwards or time-reversed causation. (Mackie, 1974/1980, p. 52)

This quote should be treated with some care. Mackie does not believe that backward causation actually exists, but only that it is conceptually possible.[8] Since it is conceptually possible, our definition of causation should not rule it out from the start. Hence, Mackie's theory of causation is time independent.

2.4.2 Salmon and Dowe's Theories of Causal Processes

Inspired by previous process views on causality, Wesley Salmon developed his own influential probabilistic theory of causal processes. His account would later be subject to some modifications to accommodate the criticism of, among others, Phil Dowe (see Salmon 1984, 1997; and Dowe 1992, 2000).

Salmon's account is built around the notion of causal processes, which are defined as spatiotemporally continuous entities with the capacity to transmit "information, structure and causal influence" (Salmon, 1994, p. 303; quoted from Galavotti, 2018, §5.5). Continuous processes form the connections between causes and effects, providing the links through which causality is propagated. To Salmon, these causal processes are "the kinds of causal connections Hume sought but was unable to find," holding that "such connections do not violate Hume's strictures against mysterious powers" (Salmon, 1990, p. 8; quoted from Galavotti, 2018, §5.5).

In defining this causal propagation through processes, Salmon was heavily inspired by both Reichenbach and his concept of *mark transmission*,[9] and Bertrand Russell and his *at–at theory* of causal propagation (Galavotti, 2018, §5.5.1-§5.5.2). From Reichenbach he borrowed the idea that causal processes, as opposed to non-causal processes, can be distinguished by their ability to transmit marks:

---

[8] In the preface to the paperback edition he writes: "My discussion [...] of backward (time-reversed) causation has often been misunderstood. I do not suggest, and I do not believe, that backward causation occurs, but only that it is conceptually possible" (Mackie, 1974/1980, p. xiv).

[9] Reichenbach supervised Salmon's Ph.D. dissertation.

A process is causal if it is *capable* of transmitting a mark, whether or not it is *actually* transmitting one. The fact that it has the capacity to transmit a mark is merely a *symptom* of the fact that it is actually transmitting something else. That other something I described as information, structure, and causal influence. (Salmon, 1994, p. 303, original italics; quoted from Galavotti, 2018, §5.5.1)

Russell's at–at theory provided a second source of inspiration. Salmon saw it as a way to account for the theory of mark transmission without introducing the mysterious causal powers warned against by Hume. In his own words: "to move from A to B is simply to occupy the intervening points at the intervening instants. It consists in being *at* particular points of space *at* corresponding moments" (Salmon, 1984, p. 153; quoted from Galavotti, 2018, §5.5.2). In the context of causal processes the at–at theory implies that when a mark is transmitted from a given point A in a process to some point B in the same process, that mark appears at each point between A and B.

Salmon's ideas drew some influential criticism, among them those espoused by Phil Dowe. Dowe suggested that causal processes should actually be defined in terms of *conserved quantities* instead of mark transmission. These suggested modifications would, for a large part,[10] be adopted by the later Salmon. Dowe's process theory of causality in terms of conserved quantities is based on the following two definitions:

Definition 1. A causal interaction is an intersection of world lines which involves exchange of a conserved quantity.
Definition 2. A causal process is a world line of an object which manifests a conserved quantity. (Dowe, 1992, p. 210; quoted from Galavotti, 2018, §5.5.3)

A *world line* is the path an object traces in four-dimensional spacetime, or in Dowe's words: "the collection of points on a space-time (Minkowski) diagram which represents the history of an object" (Dowe, 1992, p. 210; quoted from Galavotti, 2018, §5.5.3). A *conserved quantity* is defined as "any quantity universally conserved according to current scientific theories" (Dowe, 1992, p. 210; quoted from Galavotti, 2018, §5.5.3), such as charge, angular momentum, mass, etc.

Neither Salmon's account nor Dowe's proposed changes make any explicit reference to the primacy of either time or causation. (Admittedly, the concept of a world line is defined in terms of space-time, but the theories of Salmon and Dowe do not presuppose that causes precede their effects.) Salmon was inspired by Reichenbach's concept of mark transmission, but unlike Reichenbach he did not search to explain time through causation. Both Salmon's and

---

[10] Salmon actually accepted a slightly modified version of Dowe's proposal, but the specific differences matter little to our current effort.

Dowe's may thus be considered time-independent accounts of causality. This is borne out by their own work.

Let us start with Dowe:

> According to the theory argued for here, the direction of a causal process is given by the direction of an open conjunctive fork part-constituted by that process; or, if there is no such conjunctive fork, by the direction of the majority of open forks contained in the net in which the process is found. This allows backward causation where an open fork points to the past, given that the majority of forks point to the future. It also defines a direction for all processes. (Dowe, 2000, pp. 209–210)

It is clear that Dowe is envisioning a time-independent account of causation here, in which the causal direction is not derived from any temporal direction.

Dowe attributes a similar quest for a time-independent account of causation to Salmon (focusing on his 1984 book):

> [Salmon's] broad objective is to offer a theory that is consistent with the following assumptions: [...] (iv) the theory should be (in principle) time-independent so that it is consistent with a causal theory of time [...]. (Dowe, 2000, p. 66)

A decade after the publication of said book, Salmon wrote the following:

> Dowe distinguishes two types of asymmetry, causal and temporal. I agree that the two need not coincide. It is not a priori impossible for some causal processes to transpire in a direction contrary to the direction of time [...]. (Salmon, 1997, p. 462)

Again, this supports our classification of Salmon's view as a time-independent one.[11]

2.4.3 Menzies and Price's Agency Theory

Agency theories of causation, which assign a central role to human agency, start out from the idea that causal relationships are relationships that can be used for manipulation and control. One version is provided by Peter Menzies and Huw Price (1993). Price (1991, 1992, 1996) has had a continual involvement in its defense. In its most basic assertion, their theory states:

> An event A is a cause of a distinct event B just in case bringing about the occurrence of A would be an effective means by which a free agent could bring about the occurrence of B. (Menzies and Price, 1993, p. 187)

---

[11] Admittedly, slightly later in that text Salmon writes: "Like Dowe, I want to adhere to a *causal theory of time*, and this commitment precludes defining causal direction in terms of temporal direction" (Salmon, 1997, p. 463, italics added). Yet, we contend, the textual evidence fits the time-independent classification better than the causality-first one.

There are two notions in this statement that need to be unpacked (see also Illari and Russo, 2014, pp. 181–183 and Woodward, 2016, §2). First, there is the 'bringing about of occurrences'. Such a statement seemingly faces a problem of circularity (Menzies and Price, 1993, §4). The notion in question is essentially equivalent to 'causing the occurrence' as a means–end relationship. Agency theories are expected to provide a reductive analysis of causation, but the above assertion only seems to state that the causal notions are related. Menzies and Price, however, claim that there is nothing circular about their theory. In fact, circularity is avoided by appealing to the experience of acting as an agent that is independent of our grasp of the general notion of causation. In their own words:

> The basic premise is that from an early age, we all have direct experience of acting as agents. That is, we have direct experience not merely of the Humean succession of events in the external world, but of a very special class of such successions: those in which the earlier event is an action of our own, performed in circumstances in which we both desire the later event, and believe that it is more probable given the act in question than it would be otherwise. To put it more simply, we all have direct personal experience of doing one thing and thence achieving another. [. . .] It is this common and commonplace experience that licenses what amounts to an ostensive definition of the notion of 'bringing about'. In other words, these cases provide direct non-linguistic acquaintance with the concept of bringing about an event; acquaintance which does not depend on prior acquisition of any causal notion. An agency theory thus escapes the threat of circularity. (Menzies and Price, 1993, pp. 194–195)

(As correctly remarked by a reviewer, the temporal language in this quote, e.g., "earlier," "later," "thence," seems to run counter to our classification of Menzies and Price's account as time independent. We will address this issue at the end of this section.)

Answering the threat of circularity does, however, reveal another problem for Menzies and Price's account. Seeing as an agency theory is necessarily tied to the personal experience of a free agent, it becomes impossible to account for those causes that are unmanipulable, i.e. cases that do not (directly) involve an agent (Menzies and Price, 1993, §5). Their own example concerns the 1989 San Francisco earthquake, supposedly caused by friction between continental plates. It is clear that there is no human agent involved in such a case. Their argument is centered on the idea that it is possible to infer the existence of causal relations from another situation involving a pair of events that resembles the given situation with respect to its intrinsic features. In the case of the San Francisco earthquake, we could suppose that seismologists could create a model, involving artificial simulations of the movement of continental plates, which they could then use to surmise the appropriate causal claims.

Returning to the original assertion, there is another notion that needs to be explicated, namely 'effective means'. Menzies and Price (1993, §3) suggest that this idea of 'effective means' can be explained through decision theory (see also Illari and Russo, 2014, pp. 181–183). Decision theory is a way of spelling out, in probabilistic terms, what rational agents would do in a given situation to get their desired outcome. Applying this to agency theory, they introduce so-termed agent probabilities, which are "conditional probabilities, assessed from the agent's perspective under the supposition that the antecedent condition is realized *ab initio*, as a free act of the agent concerned" (Menzies and Price, 1993, p. 190, original italics). This addition of decision theory and agent probabilities introduces probability into the agency theorist's tool-box and with it stresses the central role of agency even more, as only agents are able to assign agent probabilities. Causing some event can thus be equated to an agent making the event more probable.

In two earlier papers, Price (1991, 1992) already explicitly claimed that (i) the agency account is time independent and that (ii) time-independent accounts are to be favored over time-first accounts. The introduction of agency by itself removes the need for a restriction that causes must precede their effects, such as Hume's second clause in Section 2.2.1, when we wish to account for the asymmetry of causation.

So the agent's perspective guarantees causal asymmetry, and has the consequent advantage of not excluding prematurely the possibilities of simultaneous and backward causation. (Price, 1991, p. 170)

In short, Menzies and Price's account of causation is time independent. It is incumbent upon the actions of an agent, which guarantees causal asymmetry. Similarly, the arrow of time is something experienced by an agent. An earlier event in time such as the pressing of an accelerator pedal will be chronologically followed by an acceleration of the car. This does not mean that causation merely follows the arrow of time, but rather that actions of agents in the world constitute the nature of both.

Caveat: We should add that in his 1992 paper on "Agency and Causal Asymmetry," Price comes close to a causality-first account of time (see also Price, 1996). At the beginning of that paper, he writes that requiring that causes must chronologically precede their effects, as Hume did, "precludes the project, *attractive to many*, of explicating temporal order in terms of causal order" (Price, 1992, p. 501, italics added). Towards the end he concludes that his agency theory

[...] is particularly well placed to explain the nature of causal asymmetry, and its prevailing orientation in time. For it is able to say that the asymmetry of causation simply reflects (or better, perhaps, projects) that of the means–end relation. Causes are potential means, on this view, and effects their potential ends. The origins of causal asymmetry thus lie in our experience of doing one thing to achieve another; in the fact that in the circumstances in which this is possible, we cannot reverse the order of things, bringing about the second state of affairs in order to achieve the first. This gives us the causal arrow, *the characteristic alignment of which with the temporal arrow then follows from the fact that it is normally impossible to achieve an earlier end by bringing about a later means.* (Price, 1992, p. 515, italics added)

The 'normally' in this quote is a particularly important qualifier, as it allows Price to keep his options open concerning the direction of time and the possibility of backward or simultaneous causation. He agrees that 'characteristically' there is a clear temporal orientation, one that, equivalent to the causal asymmetry, is guided by the means–end relationship that is a hallmark of an agency view of causation. But this does not mean that cases featuring backward causation could not possibly exist. Price (1992, footnote 16) refers to Dummett (1964) as a philosopher who defends the conceptual possibility of backward causation, and he even suggests that quantum mechanics might be able to provide actual examples.

2.4.4 Lewis's Counterfactual Account of Causation

As we have seen, David Hume offered two non-equivalent definitions of causation. We have already discussed the first definition in Section 2.2.1. Now we mention the two together. He wrote:

[...] we may define a cause to be *an object followed by another, and where all the objects, similar to the first, are followed by objects similar to the second. Or,* in other words, *where, if the first object had not been, the second never had existed.* (Hume, 1748, section VII; as quoted by Lewis, 1973a, p. 556)

The first definition is the regularity view discussed above. The second defines causality in counterfactual terms: if the cause had not been, the effect never had existed. Contrary to what Hume suggested, these definitions are not equivalent – the phrase "in other words" is misleading.

Hume's counterfactual definition served as a source of inspiration for David Lewis (1973a). Simplifying matters,[12] an event e *causally depends* on another event c if and only if the following two counterfactuals are true:[13]

- If c would occur then e would occur.
- If c would not occur then e would not occur.

[12] Lewis's exact formulation is not in terms of the events c and e but in terms of the propositions which state that c and e occur. This difference does not matter for our purposes.
[13] Lewis has offered an elaborate account of counterfactuals in a book published the same year (Lewis, 1973b).

The lighting of a match causally depends on its being struck (in suitable circumstances) because if the match would be struck, it would light, and if it would not be struck, it would not light.

Lewis then goes on to define *causation* in terms of causal dependence:

> Let $c, d, e, \ldots$ be a finite sequence of actual particular events such that d depends causally on c, e on d, and so on throughout. Then this sequence is a *causal chain*. Finally, one event is a *cause* of another iff there exists a causal chain leading from the first to the second. This completes my counterfactual analysis of causation. (Lewis, 1973a, p. 563, original italics)

The striking of the match is a cause of the kettle boiling since there is a chain of events linking the two (the match being struck, the match lighting, the gas stove lighting, the water in the kettle heating up) where each event causally depends on the previous one.

Lewis explicitly rejects the option of building temporal precedence into his definition of causation, even though that would help to solve a problem - the 'problem of effects' – which his counterfactual analysis faces. The problem of effects is the following:

> Suppose that c causes a subsequent event e, and that e does not also cause c. (I do not rule out closed causal loops a priori, but this case is not to be one.) Suppose further that, given the laws and some of the actual circumstances, c could not have failed to cause e. It seems to follow that if the effect e had not occurred, then its cause c would not have occurred. We have a spurious reverse causal dependence of c on e, contradicting our supposition that e did not cause c. (Lewis, 1973a, pp. 565–566)

This problem should not be solved by opting for a time-first account, Lewis writes:

> One might be tempted to solve the problem of effects by brute force: insert into the analysis a stipulation that a cause must always precede its effect [...]. I reject this solution. (1) It is worthless against the closely related problem of epiphenomena [...].[14] (2) It rejects a priori certain legitimate physical hypotheses that posit backward or simultaneous causation. (3) It trivializes any theory that seeks to define the forward direction of time as the predominant direction of causation [i.e. any causality-first account of time]. (Lewis, 1973a, p. 566)

This quote illustrates two important reasons, shared by several other philosophers discussed in this Section 2.4, for rejecting a time-first approach to

---

[14] The problem of epiphenoma is that on Lewis's counterfactual analysis, e might mistakenly be considered a cause of f if e and f are effects of a common cause c.

causality. In Lewis's case, these reasons could bring their weight to bear since he had a different solution to the problem of effects.[15] In short, Lewis's is a time-independent account of causation (which is sympathetic to the development of a causality-first account of time).

2.4.5 James Woodward's Interventionist Theory of Causation

The manipulationist or interventionist theory of causation proposed by Jim Woodward is currently very influential. The account, as its name suggests, bases the notion of causality on manipulability. According to this view, "causal and explanatory relationships are relationships that are potentially exploitable for purposes of manipulation and control" (Woodward, 2003, p. v).

Woodward's theory draws heavily from scientific practice. More specifically, it incorporates ideas on experimental design from social scientific and biomedical contexts and tries to show how they may be generalized to other areas of science, and ultimately inform a working account of causation. In its most basic formal iteration, the interventionist theory claims that:

> C is a genuine cause of E if, given the appropriate background conditions, there is a possible manipulation of the cause C such that this is also a way of manipulating or changing the effect E. (Woodward, 2003, p. 16)

What this essentially means is that if someone 'wiggles' C, they will also 'wiggle' E. Examples are quite easily found. Flipping a switch will turn a light on or off, ingesting more calories will increase your weight, or taking some aspirin will reduce your headache. Understood in this way, manipulationism is a very intuitive way of reasoning about causality.

The manipulations responsible for changing cause-variables, and in turn modifying the corresponding effect-variables, are more stringently defined by Woodward in terms of interventions (I). Interventions are changes or manipulations to a cause C performed in such a way that if any change occurs in the effect E, it occurs only because of E's relationship to C. To that effect, they must satisfy a number of conditions.[16]

An intervention on a variable X is always defined with respect to a second variable Y. "There is no such thing as an intervention on X simpliciter" (Woodward, 2003, p. 103). It is important to note that it is not necessary for the manipulations to be carried out by human agents to qualify as proper interventions. Woodward's account, contrary to the more traditional agency theories (e.g., von Wright, 1971; Menzies and Price, 1993), is non-anthropomorphic.

---

[15] For Lewis's solution to the problem of effects, see pp. 566–567 of his article.

It is Woodward's claim that, in order for C to affect a change in E there must exist a *possible* intervention on C with respect to E. 'Possible' is here taken to mean that it is only required that interventions be constrained under logical or metaphysical possibility, rather than some practical, physical or nomological possibility. Hence one may conceive an intervention on the distance of the moon with respect to the tides on earth, or on plate tectonics with respect to the 1989 earthquake in San Francisco. As long as manipulations satisfy Woodward's conditions (see footnote 16), which say nothing about human agency, they may qualify as proper interventions.

Using the notion of an intervention,[16] Woodward defines several types of causal relations. Here is one example:

> A necessary and sufficient condition for X to be a *(type-level) direct cause* of Y with respect to a variable set V is that there be a possible intervention on X that will change Y or the probability distribution of Y when one holds fixed at some value all other variables Z_i in V. (Woodward, 2003, p. 59, italics added)

The conditions described in footnote 16 ensure that there are no spurious correlations or confounding variables in play, when assessing the causal efficacy of a relation. Consider for example the correlations between smoking, yellow fingers and lung cancer (see Illari and Russo, 2014, pp. 100–101). Even though there is a strong correlation between yellow fingers and lung cancer, intervening on yellow fingers, by offering some sort of state-sponsored hand sanitizer, will not affect the incidence of lung cancer. Properly intervening on smoking on the other hand will influence lung cancer incidence in the population. Interventions on correctly isolated change-relations allow us to conclude that smoking causes lung cancer, whereas yellow fingers do not.

---

[16] Interventions are defined in terms of an intervention variable. I is an intervention variable for C with respect to E if and only if it meets the following conditions (Woodward, 2003, p. 98, notation slightly changed):

I1. I causes C.

I2. I acts as a switch for all the other variables that cause C. That is, certain values of I are such that when I attains those values, C ceases to depend on the values of other variables that cause C and instead depends only on the value taken by I.

I3. Any directed path from I to E goes through C. That is, I does not directly cause E and is not a cause of any causes of E that are distinct from C except, of course, for those causes of E, if any, that are built into the $I - C - E$ connection itself; that is, except for (a) any causes of E that are effects of C (i.e., variables that are causally between C and E) and (b) any causes of E that are between I and C and have no effect on E independently of C.

I4. I is (statistically) independent of any variable Z that causes E and that is on a directed path that does not go through C.

In essence, the above termed conditions specify that if E is affected, this is solely through the action of the cause C, triggered by the intervention I.

Woodward's account can be said to be time independent. He makes no direct reference to anything akin to temporal ordering in his definitions of causation, interventions and intervention variables. The only asymmetry Woodward is interested in, is the "underlying physical asymmetry [...], an asymmetry connected to facts about manipulation and control" (Woodward, 2003, p. 197). This does not mean, of course, that Woodward endorses the idea that effects may precede their causes. But that they should follow their causes is not part of his definitions of causation. Consider the following hypothetical example: If it were possible to intervene on a light switch at time t_2 and for that intervention to have a robust effect on a light at an earlier time t_1, this would satisfy Woodward's account for a causal relationship while also being a backward-directed relationship in time.[17]

2.5 Two Further Issues: Instantaneous and Backward Causation

In the preceding sections, we have repeatedly found references to instantaneous and backward causation. What are these 'phenomena' that seemingly contradict the common sense idea that we started from in this chapter?

2.5.1 Instantaneous Causation

Instantaneous, or synchronic, causation is a form of causation where the cause and the effect occur at the same time. Instantaneous causation was once a well-accepted notion. It is now less popular, yet quantum theory has inspired some researchers to once again consider its possibility.

Historically, instantaneous causation has been defended by philosophers from René Descartes (seventeenth century) to Immanuel Kant (eighteenth century). The latter gave a well-known example. Consider a lead ball resting on a cushion. The presence of the ball on the cushion creates a visible indentation. Cause and effect seem to occur simultaneously in this case.[18] (Kant, 1781/1998, p. A203)

The above example, however, poses a problem from a physical standpoint. For a cause and an effect to happen simultaneously, causal influence would have to be propagated instantaneously. Intuitively, this seems to run counter to some basic assumptions about the physical world, at least if one thinks of causality in terms of mark transmission (cf. Reichenbach, Salmon, and Dowe).

[17] We express our thanks to the reviewer who supplied this concrete example.
[18] Similar arguments have been used by theists to ground divine causation.

The special theory of relativity forbids any instantaneous, or 'spooky', action at a distance. In this view, a truly simultaneous case of cause and effect would mean that causal influence was somehow transmitted faster than the speed of light. It is then fair to question if such a thing as instantaneous causation could even be possible at all.

Certain advances in physics, however, have suggested possible cases of instantaneous causality. Hence the issue is divisive within the field of physics. One such case results from the so-termed EPR-paradox (Tooley, 1997, pp. 354–359). Essentially, the paradox involves two particles that are quantum entangled. Under the standard interpretation of quantum mechanics, each particle is individually in an uncertain state until it is measured, at which point the state of that particle becomes certain. At that exact same moment, the other particle's state also becomes certain. This is paradoxical as it seemingly involves some sort of superluminal transfer between the two particles. This instantaneous action at a distance runs counter to Einstein's special theory of relativity.[19]

A second example is the simple pendulum: the length l of the pendulum is a cause of its period T.[20] Leuridan (2012) takes this causal relation to be a possible case of a synchronic invariant relation, or an instance of instantaneous causation, seeing as it could be argued that the change in T does not follow the change of length in time.

A third case of instantaneous causation is presented by Huemer and Kovitz (2003):

> The Lorentz equation in the theory of electromagnetism provides another example of simultaneous causation: a body with charge q moving at velocity v through electric and magnetic fields experiences a force given by
>
> $$\vec{F} = q\vec{E} + q\vec{v} \times \vec{B}$$
>
> where $\vec{E}$ and $\vec{B}$ are electric and magnetic field vectors at the body's current location in space. The vectors $\vec{E}$, $\vec{v}$ and $\vec{B}$ will typically vary over time, along with $\vec{F}$, and the value of $\vec{F}$ at any given time is determined by the values of $\vec{E}$, $\vec{v}$ and $\vec{B}$ *at that time*. Bodies always experience the effect of the electromagnetic field at their location in space and time. (Huemer and Kovitz, 2003, p. 559, original italics, mathematical notation changed)

The charge's acceleration occurs simultaneously with the application of the electric and magnetic fields.

---

[19] Although some experiments on quantum entanglement do seem to support faster- than-light information transfer, and Bell's theorem suggests non-locality, the matter is still very much an open question.

[20] Cf. any Woodwardian intervention on l will cause a change in T; see Woodward (2003, p. 197) and Leuridan (2012, pp. 419–420).

From a philosophical standpoint, instantaneous causation is equally divisive. First, any account of causation that subscribes to the assumption that causes must precede their effects, i.e. any time-first account, will naturally be opposed to instantaneous causation. We have discussed Hume's and Mill's regularity accounts as examples.

Second, time-independent accounts, as their name suggests, do not rely on temporal priority. This would suggest that they are able to allow for instantaneous causation. In the case of some philosophers (Lewis, Price), we have established that they have consciously made room for such a possibility.

Third, a causality-first account such as Reichenbach's uses causality to enforce a clear direction in time. It stands to reason that instantaneous causality would be problematic for Reichenbach, yet nowhere in *The Direction of Time* does he explicitly exclude simultaneous causation. (On pp. 39–40 he comes the closest to excluding it.)

It is an open question whether this resistance to instantaneous causation would apply to other causality-first accounts of time as well.

In general, instantaneous causation is not really given much serious consideration in the philosophical literature. This may, in part, be due to the common sense intuition that we started this chapter with; common sense seems to dictate that causes must precede their effects. There are of course exceptions to this rule, yet real-world examples of simultaneous causation, such as those given above, are rather rare.

In order to further illustrate the philosophical divisiveness of the issue of instantaneous causation, it is useful to distinguish another strand in the literature. Some philosophers explicitly advocate the view that causes and effects must always be simultaneous. We have already mentioned Huemer and Kovitz (2003). They base their account on certain key assumptions, such as a distinct view of the mathematical structure underlying time, that differ significantly from more standard accounts. Mumford and Anjum (2011) have similarly developed a simultaneous view of causation. They argue for *causal dispositionalism*, a theory of causation derived from an ontology of real dispositions or powers. Cause and effect must be simultaneous as, in their view, causation is the process that occurs whenever partnered powers actually produce their effect, and not before. It begins once those partners are together and ends either when the process is complete or is prematurely interrupted. Consider a billiard table. Causation does not occur until two billiard balls actually touch and it is over once they no longer do.[21]

---

[21] As Mumford and Anjum acknowledge, their simultaneous view of causation requires them "to meet the challenge of explaining how causation and causal chains can take time" (2011, p. ix), a challenge that they take up in their fifth chapter.

Theories of instantaneous causation seem to require some major paradigm shifts to ground their ideas. For example, Huemer and Kovitz warn that "[g]rasping the theory of simultaneous causation requires some cognitive shifts away from the sequential" (2003, p. 557).

2.5.2 Backward Causation

Backward causation is the idea that the effect precedes its cause in time. It can be defined more stringently as:

> [...] the idea that the temporal order of cause and effect is a mere contingent feature and that there may be cases where the cause is causally prior to its effect but where the temporal order of the cause and effect is reversed with respect to normal causation, i.e., there may be cases where the effect temporally, but not causally, precedes its cause. (Faye, 2015, Intro)

Given the common sense view that causes must precede their effects, the notion of backward causality may seem counterintuitive. It was not given any serious consideration until the late 1950s, when Michael Dummett (1964) first stated that there was no philosophical objection to effects preceding their causes. This was met with fierce criticism from his peers, which eventually led to Max Black's influential 'bilking argument' (Faye, 2015, §1). This argument claims that backward causality is both logically and conceptually impossible:

> Imagine B to be earlier than A, and let B be the alleged effect of A. Thus we assume that A causes B even though A is later than B. The idea behind the bilking argument is that whenever B has occurred, it is possible, in principle, to intervene in the course of events and prohibit A from occurring. But if this is the case, A cannot be the cause of B; hence, we cannot have backward causation. (Faye, 2015, §1)

In physics, backward causation has met with a different set of problems and considerations (Faye, 2015, §4). First, it should be noted that there do not seem to be any generally accepted empirical phenomena that demand a notion of backward causation. Second, it is worth remembering that according to many philosophers of physics, causality has no or a very limited application in physics.[22] We may add to this that our ability to recognize causal relations in physical processes is largely determined by the specific notion of causality one subscribes to (see the various theories of causation described in Sections 2.2, 2.3, and 2.4). It is nevertheless interesting to approach the possibility of backward causation from a physicist's point of view, not in the least because

---

[22] A notable exception is Frisch (2014). See also De Bal (2017).

the most fundamental laws of nature are considered time reversal invariant in the sense that our physical theories allow description of the fundamental reactions and processes in a reversed temporal order (Illari and Russo, 2014, pp. 178–179). These processes are thus said to be reversible in time (Faye, 2015, §4). Certain aspects of modern physics may also allow particles or information to travel backward in time, cf. hypothetical tachyons and/or Bell phenomena.

In philosophy, backward causation creates conceptual problems beyond just the bilking argument (Faye, 2015, §2).[23] First, we need a metaphysics capable of allowing an effect to precede its cause. Backward causation seems to require a static view on time, meaning that it would not feature any kind of coming-into-being. The future is real; it already exists, next to the past and the present, and it contains facts that can make sentences about the future true or false. A second problem, which is more directly relevant to the present chapter, is the following: "*Can the cause be distinguished from its effect so that the distinction does not depend on a temporal ordering of the events?*" (Faye, 2015, §2, original emphasis). Our typology may help to address this question in a straightforward manner: making the required distinction can be done in time-independent theories of causation and in causality-first accounts of time, but not in time-first accounts of causation.

Faye's question gives rise to a closely related issue: to what extent can the three different strands allow for backward causation? First, time-first accounts of causation necessarily exclude its possibility as they depend on a temporal ordering of cause (first) and effect (later). Second, though his causality-first account acknowledges the seemingly time-reversible nature of micro-processes, Reichenbach attempted to explain the direction, and hence asymmetry, of time in terms of the purported asymmetry of causality. This seems to exclude the possibility of backward causation – at least at the macro-level. Third, for the proponent of backward causation, time-independent accounts of causality may fare better, as they tend to make no explicit reference to either the nature or the direction of time. As we have seen, many time-independent theorists have explicitly left the door open to backward causation (Mackie, Price, Salmon, Dowe, and Lewis). As far as we know, Woodward has not touched the subject directly. We can, however, tentatively suppose that he could allow for backward causation on the condition that e.g. interventions on later events may be used to change earlier events.

---

[23] Faye discusses four conceptual problems. We leave the second and last one aside. Apart from these conceptual problems, backward causation also seems to give rise to a number of logical paradoxes (e.g. bootstrap, consistency, Newcomb) (Faye, 2015, §3).

2.6 Cards on the Table

So far we have only offered a typology of theories of causation, without showing any preference for one or the other strand. Let us lay our cards on the table now. In defense of Hume and Mill, one could say that opting for a time-first approach was very sensible at the time. Few if any convincing cases of instantaneous (let alone backward) causation were known and their time-first approach was well in line with the common sense intuition that causes must precede their effects. Since then, however, the empirical landscape has changed.

Both Reichenbach's project (to give an account of the direction of time) and his take on it (to start from the purported asymmetry of causation) were highly valuable in their own right. It is not, however, our project. We are primarily interested in (i) definitional or conceptual questions, in (ii) epistemological or methodological questions and in (iii) practical questions concerning causality, not time.

In our search for answers to these questions, we are most in favor of time-independent approaches to causation. We endorse the three reasons for preferring time-independent over time-first accounts listed at the beginning of Section 2.4. First, they are conceptually less costly (subject to our remark in footnote 6). Second, they leave open the possibility of using one's account of causation to establish a causal account of time – even if that is not our own project. Third, they are less vulnerable to the possibility of backward or instantaneous causation which, given the aforementioned change in the empirical landscape, has become more scientifically palatable.

Acknowledgments

The research for this chapter was supported by the Research Foundation Flanders (FWO) through research project G056616N. We would like to thank Samantha Kleinberg and the reviewers for their constructive and helpful comments.

3

The Direction of Causation

Phil Dowe

Suppose C and E are causally linked in the sense that either C causes E or E causes C. What makes it the case that C causes E rather than that E causes C? Being overweight and getting inadequate exercise seem to be causally linked, but which causes which? We can ask this for both type and token events. Type events, or properties, are general – as in the example I just gave. Token events, or property instantiations, are specific instances. Thus, my being overweight and my lack of exercise are causally linked, but which caused which? (For types we will use upper case letters, for tokens lower case.) My question is not the epistemological question 'how do we find out which caused which?', but the metaphysical question 'what makes it the case that the causal arrow points a particular way?'. This is the topic I call the direction of causation.

This should not be confused with certain questions about the asymmetry of causation. One thing that term can mean concerns a logical property of a relation – we say a relation R is asymmetric iff 'a's being related by R to b' entails it is not the case that b is related by R to a; e.g. a's being smaller than b entails it is not the case that b is smaller than a, where the asymmetry is a logical property of the relation 'smaller than'. I don't think causation is asymmetric in that sense; e.g. it may be that being overweight and inadequately exercising cause each other at the type level, as is the case for any feedback loop. Neither is causation symmetric, where 'symmetric' refers to the logical property of a relation such that 'a's being related by R to b' entails 'b is related by R to a'. 'Is equal to', 'is a classmate of', and 'are causally linked', are symmetric relations. But even if being overweight and inadequate exercise cause each other, smoking and cancer do not. So 'a causes b' entails nothing about whether b causes a; that's contingent, so we should say causation is a non-symmetric relation. In my view this is also so for token causation because

37

there can also be token causal loops, e.g. if there are closed time-like curves. Another thing 'asymmetry of causation' can mean concerns the orientation in time of causes and effects. If all causes occur before their effects, or most do, then we have an asymmetry of causation in time. No doubt causation is asymmetric in that sense, but whether that's true is not the same question as what makes it the case that the causal arrow points a particular way. In posing the question of the direction of causation, I said nothing about time. Nevertheless, any assertion of an asymmetry of causation in time assumes an answer to the question of the direction of causation.

Now let's add a constraint. David Lewis says:

> Careful readers have thought they could make sense of stories of time travel
> [...] speculative physicists have given serious consideration to tachyons,
> advanced potentials, and cosmological models with closed timelike curves.
> Most or all of these phenomena would involve special exceptions to the normal
> asymmetry of counterfactual dependence. It will not do to declare them
> impossible a priori. (Lewis, 1979, 464)

I agree with this sentiment, indeed it will not do to declare them impossible a priori. I'm not going to argue for the constraint; rather I will assume it. My goal in this chapter is to explore its implications. The constraint places quite severe restrictions on our task of saying what the direction of causation consists in. First and foremost it entails that the direction of causation cannot be fixed by the time order of the events. This rules out an easy and common answer to our question concerning the direction of causation, the answer given for example by David Hume, who said 'we may define a cause to be an object, followed by another, and where all the objects, similar to the first, are followed by objects similar to the second' (Hume, 1748, ch. 7) On this 'regularity' account, c and e are causally linked if c is before e and there's constant conjunction of C and E, where the C-type events are always before the E-type events. The direction of causation is given by the order in time. Our constraint tells us to reject this because that would be to declare backwards in time causation impossible a priori. Some illustrations will help set the scene.

Take one of Lewis's examples, the Wheeler and Feynman (1949) advanced potential. Wiggle a charge and you create an electromagnetic field, described by Maxwell's equations, which radiates into the future. To picture this look at the WiFi symbol on your computer and imagine the wiggled charge at the point. Maxwell's equations admit two solutions, one normally disregarded as unphysical (thereby obviating the time-symmetry of the laws of physics). To overcome various problems with that approach, Wheeler and Feynman proposed that we don't disregard the advanced solution and they developed electrodynamics on that assumption, thereby reinstating the time-symmetry of the laws of physics (for other advantages, see de Souza and De Luca (2015)). On their account, if you wiggle a charge you create an electromagnetic field which radiates symmetrically into the future and the past. To picture this, look at the WiFi symbol and imagine in addition a mirror image down from the point. When you wiggle a charged particle (c) you can thereby cause other charged particles to move: in the future, say e, and also in the past, say a. Then we have the causal schema $a \leftarrow c \rightarrow e$ with the time order a, c, e. Speculative physicists have given serious consideration to advanced potentials, and it will not do to declare them impossible a priori. This entails that the causal direction cannot be defined in terms of the time order. Notice that I've analysed the Wheeler and Feynman theory in terms of causation, and in doing so I've assumed that there is some satisfactory answer to the question of the direction of causation. More specifically, I said, wiggle this, and so that wiggles. Here I was drawing on an Interventionist account of causation, and my task below will be to explore *whether* intervention fixes the direction of causation.

As a second example, consider retrocausality in quantum mechanics. In the standard Bell experiment we let two electrons interact then measure their spin when they're separated. The quantum state produced by the interaction is a superposition of vectors corresponding to the possible measurement outcomes. This is an entangled state – it produces a particularly strong correlation between the measurement outcomes A and B. One way of putting the challenge of entanglement is to say it violates the Causal Markov Condition (see below) in that, conditional on the quantum state (the obvious common cause) A and B remain correlated. Retrocausality is a way to explain this. One version of retrocausality developed by John Cramer takes advantage of the Wheeler–Feynman advanced potentials (Cramer, 1980). Another version can be schematised as follows: suppose the quantum state is incomplete and in fact the complete state λ immediately after interaction takes values that determine the measurement outcomes A and B. For a given preparation, λ is probabilistically determined by the pair of measurements setting S, T (Evans, 2017, 750–752). This causal model does not violate the Causal Markov Condition since conditional on λ as common cause, A and B are not correlated. For spin measurements on the pair of electrons, the settings are angles of the Stern–Gerlach devices used to register the spin outcomes. In contemporary experiments these settings can be determined randomly in the forward light cone of the preparation interaction. Thus, the retrocausal part of the schema is: $S \rightarrow \lambda \leftarrow T$ where λ occurs before S, T. Again, if we defined causal direction in terms of the time order this explanation would not be available.

Lewis wants to allow these kinds of backwards causation. On his account of causation, c causes e if the counterfactual 'had c not occurred e would not have occurred' is true. That counterfactual is true if at the closest possible worlds where c doesn't occur, e doesn't occur (Lewis, 1973a). Which of the closest possible worlds where c doesn't occur are 'closest' is given by a similarity measure that ranks worlds by balancing two factors: the extent to which they contain local sequences of events that are a violation of the laws of the actual world ('miracles'), and the extent to which they contain regions of perfect match to the actual world with respect to particular matters of fact. The first factor subdivides into 'small miracles' and 'large miracles,' the latter being collections of diverse and widespread cases of the former. His similarity measure says it is of first importance to minimise large miracles, of second importance to maximise regions of perfect match, and of third importance to minimise small miracles (Lewis, 1979, 472). In typical cases of causation the closest worlds will be ones that perfectly match the actual world up until a time just before the time of the putative cause c, then by a small miracle no event of type C occurs, and thereafter diverge from the actual world. If in all such closest worlds there is no event of type E, then c causes e. One might wonder why the following type of world is not closer still: perfect match to the actual world up until a time just before the time of c, then by a small miracle no event of type C occurs, then a further small 'reconvergence' miracle to return to perfect match, and thereafter perfect match. After all, it's worth a small miracle to buy large regions of perfect match. But given the nature of the actual world, no such worlds are among the closest, which is a good job since if they were, c would not count as a cause of e. Such worlds would contain events of type E at the time of e. Given the nature of the actual world, it normally takes a large miracle of reconvergence, and by Lewis's similarity measure gaining extra regions of perfect match is not worth a large miracle.

Why does it take a large miracle to reconverge? Because our world exhibits an 'asymmetry of overdetermination' (Lewis, 1979, 474). A determinant of a fact is a minimal set of conditions jointly sufficient, given the laws of nature, for the fact in question. If there are many such determinants in, say, the future of the fact then that fact is 'overdetermined' in the future. Now, according to Lewis our world is such that events have much more future overdetermination than past overdetermination. This is the asymmetry of overdetermination. The example Lewis gives is a spherical wave that expands outward from point source (Lewis, 1979, 475). 'Countless tiny samples' of the wave determine the source given the laws. The reverse does not normally occur in a world like ours.

And if our world did not exhibit this asymmetry in some direction then nothing would count as causation. Therefore on Lewis's account it is required, for there to be causation, that the cause exhibits an asymmetry of overdetermination in the direction of the effect. This is what makes it so that the arrow on the causal connection goes the way it does. That's his answer to our question of the direction of causation.

There are two problems with this account. The first is that Lewis can't account for the very case of backwards causation that he himself cited. On the Wheeler–Feynman electrodynamics when you wiggle a charge (c) you can thereby cause another charge (e) to move in the future, and also a third charge to move (a) in the past. This requires that c exhibits an asymmetry of overdetermination towards the future (in order for it to count as a cause of e) and an asymmetry of overdetermination towards the past (in order for it to count as a cause of a). This, of course, is impossible. The electromagnetic field radiates symmetrically into the future and the past, so the overdetermination of c is symmetric in the future and the past. There is no asymmetry of overdetermination. So what are the closest $\sim c$ worlds on Lewis's similarity measure? Since the absence of large miracles trumps a gain in regions of perfect match, the closest $\sim c$ worlds will be worlds with no perfect match at all. Some of those worlds will contain a and/or e. The problem here is that Lewis's similarity measure no longer in effect 'holds fixed the background,' which is needed because probably nothing is a nomologically necessary condition in *any* physically possible circumstances. So nothing's stopping a and e from occurring by some other means, without c. Therefore, contrary to advertisement, Lewis's theory does not count advanced potentials as a case of causation.

A second, but related problem is completely general and concerns Lewis's idea of a determinant. As Frisch (2014, 204–205) points out, the existence of a tiny wave sample does not, given the laws, imply the presence of a source – it's also compatible with the assumption that the radiation is source free. To entail the presence of a source, you would need to specify a fully specified state over the light cone. This applies not only to radiation but completely generally. Even for a single particle, a later local state of the particle does not entail, given the laws, the earlier state of the particle. To have that entailment you would need a complete specification of the future light cone to hold fixed that there are no other particles interacting.

Perhaps the direction of causation is determined not by such a pattern of determinants, but by some pattern of correlations. Dan Hausman has given such an account, roughly:

... causal priority consists in the causal connection among effects of a common cause and the causal independence among the causes of a given effect. (Hausman, 1998, 55)

Hausman calls this independence. Causal connection or causal dependence means one causes the other or they have a common cause, and causal independence means not causally connected. Take the type schema: $D \leftarrow C \rightarrow E \leftarrow A$. Call this Scheme 1. C and A are independent, while D and E are dependent. Then the causal arrow points from C to E in virtue of the fact that C's effects are dependent and E's causes are independent. There is no appeal to time order – it makes no difference whether E happens to be before C.

Does causal modelling not say the same thing? By 'causal modelling' I mean specifically the algorithms of Spirtes et al. (2000) and Pearl (2000) for generating causal structure from data presupposing certain principles, in particular for our purposes the Causal Markov Condition and the Faithfulness Condition. The Causal Markov Condition states that a variable in a causal model is independent of its non-descendants conditional on its parents, where a descendant of a variable is any variable causally downstream and the parents of a variable are its immediate causes. Suppose, along with the DAG ('directed acyclical graph' – the graphical representation of causal structure) for Scheme 1, we have the statistical data:

$$A \perp\!\!\!\perp C, D \perp\!\!\!\perp E|C$$

$$A \not\!\perp\!\!\!\perp E, D \not\!\perp\!\!\!\perp E, C \not\!\perp\!\!\!\perp D, D \not\!\perp\!\!\!\perp W$$

where $\perp\!\!\!\perp$ represents statistical independence, i.e. $A \perp\!\!\!\perp C$ means $P(A \wedge C) = P(A)P(C)$; and '$\not\!\perp\!\!\!\perp$' represents statistical dependence, i.e. '$A \not\!\perp\!\!\!\perp E$' means $P(A \wedge E) \neq P(A)P(E)$. The model for Scheme 1 respects the Causal Markov Condition simply because D and E are uncorrelated conditional on C; and C, which has no parents, is uncorrelated with A and vice versa. For Hausman, for C to cause E there must be some cause of E that is not causally connected to C, and there is – A. But without A (or some other cause of E that is not causally connected to C) there would be nothing that makes it the case that C causes E. Pretty much the same can be said for causal modelling. The DAG for Scheme 1 is not the only causal model that respects the Causal Markov Condition for this data: turn the $D - C$ arrow around and you get another Markov model, these models are 'Markov equivalent'. To take the obvious example of Markov equivalence, the data: $D \not\!\perp\!\!\!\perp C, C \not\!\perp\!\!\!\perp E, D \not\!\perp\!\!\!\perp E, D \perp\!\!\!\perp E|C$ is compatible with

$$D \rightarrow C \rightarrow E,$$
$$D \leftarrow C \leftarrow E \text{ and}$$
$$D \leftarrow C \rightarrow E.$$

So with independent causes of E, causal modelling determines the arrow from C to E, without it it doesn't. Nevertheless there is this difference: causal modelling says we don't know which of the Markov equivalent models is the case, whereas Hausman says none of them involve causation. On Hausman's account, defining the direction of causation between two events requires at least four events.

That's one way Hausman's account is different to causal modelling. There's also a more significant difference. I haven't said that for Hausman 'causally connected', means correlated, or that 'causally independent' means uncorrelated. Hausman thinks that typically causally connected events are correlated and causally independent events are not; but not always. He calls it the Operationalizing Assumption, but doesn't accept that it's a definition. Why not always? Because variables not causally connected can be correlated, and causally connected variables can be uncorrelated. The counterexamples are well known: bread price increases in England are correlated with but not causally connected to rises in sea level in Venice. And causally connected variables can be uncorrelated, specifically in unfaithful causal models. The Faithfulness Condition says the only independencies are those generated by the Causal Markov Condition. For example, kookaburras kill snakes and mice, and snakes kill mice. Suppose it happens to be that the number of mice (C) is statistically independent of the number of kookaburras (A), because any change in the number of kookaburras is balanced by a corresponding converse change in the number of snakes (B). Thus, we have $C \not\perp B, A \not\perp B$, and $A \perp C$. But $A \perp C$ is not generated by the Causal Markov Condition for the only reasonable DAG, i.e. where A points to B and C, and B points to C. That causal model is therefore unfaithful. For this reason, Hausman thinks that causal dependence need not mean statistical dependence. Causal modelling, on the other hand, will not give us this model since it violates Faithfulness, and indeed cannot give us any causal model. Causal modellers have responded by saying this parameterisation is too unlikely, or that the independence is unstable – i.e. it will disappear on slight changes to the background conditions (Spirtes et al., 2000). But the fact remains that this data is possible and causal modelling can't give you the causal structure which, by the way, is not itself unstable.

So for Hausman, causal connection and causal independence cannot be defined in terms of statistics. He prefers to take the Operationalizing Assumption as a sketch of a definition of a theoretical relation:

> Tokens are causally connected if they bear that relation to one another that
> typically issues in probabilistic dependencies between their types. (Hausman,
> 1998, 58)

So, causal connection is a theoretical relation in the manner of Lewis's
functionalism about the mind (Lewis, 1972), where, necessarily, mental states
are whatever it is that plays a certain causal role with respect to perceptions
etc. and behaviour, and contingently the things that play this causal role
are brain states, at least for humans. Peter Menzies (1996) has applied this
approach to causation: causation is whatever it is that typically plays the
counterfactual chance-raising role. For Menzies, 'typically' is a way to avoid
counterexamples involving preemption. For Hausman, then, 'typically' is a
way to avoid counterexamples involving unfaithfulness. But neither Menzies
nor Hausman tell us what actually plays that role. In the case of functionalism,
we know exactly what is claimed about the contingent identity – mental states
are brain states. But for causation, what actually is that relation that typically
gives you the right pattern of correlations? And if we don't know what it
actually is, we can hardly say we have an account of the direction of causation.

Menzies suggests it might be energy transfer or exchange of energy in an
interaction. This might explain our counterexamples: in preemption cases there
is an actual physical process or mechanism connecting the effect to the actual
cause but not to the preempted event; in the correlated but causally independent
case there is no physical process or mechanism connecting English bread
prices and Venetian sea levels; and in the unfaithful case there is an actual
physical process or mechanism connecting the kookaburras to the mice. But
energy transfer or conserved quantity exchange can't play the role we want: to
say there's transfer of momentum from C to E implicitly assumes a temporal
order. There may be a fixed momentum at each point along the process but
what fixes it so that it's transferred from C to E rather than from E to C?
The same is true for an interaction – conservation of momentum tells you that
the total momentum before and after the interaction is constant for a closed
system, but gives no special privilege to before and after. What you need in the
latter case is something akin to molecular chaos: Maxwell and Boltzmann's
assumption that at each collision in the evolution of a system of molecules
comprising a gas the velocities of incoming molecules are not correlated in the
way that they will be after the collision – an assumption that Boltzmann used
to formulate the second law in terms of probability: a system will evolve to the
Maxwell Distribution (equilibrium). The connection with the kinetic theory of
gases is of interest, but for this to be an answer to our question we need to
know that molecular chaos is more than just an assumption (see Price (1996)),

and how the behaviour of Newtonian molecules in gases in a certain density range is relevant to my being overweight or my lack of exercise (see Uffink (2017)). That seems like a big ask.

In causal modelling Markov equivalent models are usually resolved by appealing to time order or to interventions. In the case of time order, in our example which of D, C, E comes first determines which causal model is correct. But we're constrained not to appeal to time order to fix the direction of causation. In the case of interventions, we can intervene on C and see whether D and/or E change, which will determine in our example which causal model is correct (the Do operator in Pearl (2000)). So does this give us our answer to the direction of causation?

According to Woodward's (2003) interventionism, we have causation if by intervening on the cause variable C to set it to a specific value we change the value or probability of the effect variable E. In the context of a given causal model, an intervention I must ideally meet four criteria. I must be a type-level cause of C. I must be a 'surgical intervention' that only affects E via C; it must be 'arrow breaking' in that it disrupts the correlation C has with any of its parents – i.e. we choose the value of I to overpower the other causes of C. And I must be exogenous – i.e. not correlated with any variable in the model other than C and its descendants. Similar concepts are used in causal modelling, e.g. Pearl (2000), this is again pretty much the 'do operator'. But Woodward's account starts with the concept of an intervention and that's a significant difference. For example, it allows him to admit at least the possibility of violations of the Causal Markov Condition and Faithfulness. Intervene on bread prices and you won't change the sea level in Venice; intervene on the kookaburra population size holding fixed the snake population and you will change the mice population.

Does interventionism give us an adequate account of the direction of causation? C causes E and not vice versa if we intervene on C and the probability distribution of E changes, while if we intervene on E the probability distribution of C is unchanged. But as Woodward explained from the start, this is circular – the definition of causation appeals to interventions and the definition of interventions appeals to causation. Therefore there's a regress – for it to be the case that C causes E it must be that the intervention I causes C; for it to be the case that I causes C it must be that some intervention, say $I*$, on I changes C and for that to be so it must be the case that $I*$ causes I, and so on in an infinite regress. This is not lost on Woodward; it's not a reductive account, he says (Woodward, 2003, 20). So interventionism doesn't tell what the direction of causation is. That question is never answered.

Can you can avoid this regress if you explicate the notion of an intervention in terms of agency or free will? Call this the agency theory:

> … an event A is a cause of a distinct event B just in case bringing about the occurrence of A would be an effective means by which a free agent could bring about the occurrence of B. (Menzies and Price, 1993, 187)

According to Menzies and Price A is an effective means by which a free agent could bring about the occurrence of B iff $P(B|A) > P(B|\sim A)$ where these are understood as 'agent probabilities', conditional probabilities where the antecedent condition is 'realized *ab initio*, as a free act of the agent concerned' (Menzies and Price, 1993, 190). The account thus distinguishes spurious from causal correlations by what interventionists call arrow breaking. If a correlation between A and B is confounded by a common cause C: $A \leftarrow C \rightarrow B$, then the overpowering of A by the free agent will leave no correlation with B.

Taken as a reductive analysis of causation, the agency theory has been the subject of many criticisms. There are cases where there is causation that makes no difference, specifically in unfaithful models (Hausman and Woodward, 1999, 534) such as our example of the kookaburras, snakes and mice. To avoid this, one would need to hold fixed certain background conditions along the lines of the interventionist definition of a cause. Next, there are the cases where the intervention itself is correlated with another relevant variable (Woodward, 2003, 240): I begin an exercise regime but I find it uncomfortable due to my weight. I decide to also go on a fad diet and over time I do lose weight. To me the fad diet is a means to the end of losing weight, but suppose in fact it's confounded by my exercising, the real cause of the weight loss. My decision might in theory break arrows leading to taking the fad diet, but it won't break arrows that don't come via the 'free act'. Hence, Woodward requires that interventions are not correlated with any variable on a causal path that does not go through A.

Some proponents of so-called agent causation think that agents are prime movers – they start new causal chains but are not caused – and that such libertarian free will solves these problems. I'll leave it to libertarians to explain how my exercising doesn't cause me to choose a fad diet. The answer has to entail that the intervention and the effect cannot have a common cause. In effect, agent causation plays a similar role to randomisation. Indeed, in quantum physics it is common to speak of 'free will variables'. Thus Wiseman defines agent causation as follows: 'If an event is statistically dependent on a freely chosen action, then that action is a cause of that event'

(Wiseman, 2014, 22). In fact what generally happens in quantum experiments is that the choice of settings are randomised ('free will variables') and the point of this is to rule out confounding. This of course is precisely what happens in randomised controlled trials in statistical sciences, and it addresses confounding of both the a–b correlation and the I–b correlation. Menzies and Price help themselves to the same strategy. The intervention is taken to arise *ab initio*. Does this mean they are committed to libertarian free will? Not at all. They take their account to be an analysis of the concept of causation rather than of causation in the world. This does raise the worry that the idea of causation is not applicable to the world because we are not in fact prime movers. Randomisation is still possible, but the connection to agents is then less clear.

There are other sorts of cases where randomisation doesn't help: 'fat handed' interventions. John wants his insomnia to stop and he thinks dealing with his depression would be an effective means, so he takes antidepressants and his insomnia is alleviated. But for depression to be a cause of insomnia it better not be that the antidepressants are confounders which alleviate depression and insomnia. If that is the case, then John is just wrong to think that countering depression is a means for alleviating insomnia. The free act might break arrows leading to depression but it won't stop the intervention causing the effect by another route. Hence, Woodward requires surgical interventions. Double blinding in statistical sciences is intended to address this problem.

I don't see why these criticisms need be fatal to the Menzies–Price agency theory, because the agency theory could simply be cast in terms of Woodward's interventionism along with its conditions for exogenous, arrow breaking and surgical interventions. Then there would be no need for a prime mover. But there is one condition in interventionism that remains a problem: the intervention variable must be a cause of the putative cause, and as we've seen this leads Woodward to agree the account is circular, and hence to deny that it's intended as a reductive analysis. This problem has often been pointed out, because it's hard to deny that an agent's bringing something about is a case of causation. So agency no more answers the question about the direction of causation than does Woodward's interventionism.

Price has disavowed the reductive analysis of Menzies and Price in favour of what he calls 'philosophical anthropology': 'the task of explaining why creatures in our situation come to speak and think in certain ways—in this case, in ways that involve causal concepts' (Price, 2017a, 75). Of course, construed this way the agency theory no longer gives an answer to our metaphysical question about the direction of causation. Nevertheless, Price's account does bear on our question. Price takes it that the direction of causation is a perspective that arises because of our asymmetric orientation as agents. This entails that there is no objective fact of the matter (independent of agents' perspectives) as to what direction the arrow is placed on a causal connection. One argument for this is based on the possibility that there could be agents who occupy a region of large-scale entropy reversal, who would see the direction of causation differently to us (Price, 2018, 83). I think this equivocates on 'the direction of causation'. For any choice of time co-ordinates, such agents would see the direction of time reversed compared to us. They would also see the direction of the temporal asymmetry of causation reversed compared to us: they would see most if not all causation going backwards. But that doesn't mean they see what constitutes the direction of causation in any sense reversed. If that were so, they might see someone's cancer causing their smoking. That would truly be perspectival. Of course, that's not what we expect of an entropy reversal region. For any of the accounts discussed above (my difficulties notwithstanding) what constitutes the direction of causation would be the same in the entropy reversal region, in the sense that it would arise equally from the asymmetry of overdetermination or the independence of causes or... The same causes would cause the same effects, type or token. There's no case based on entropy reversal for perspectivalism about the direction of causation.

Nevertheless, in summary, we have a negative result. We have not arrived at an answer to the metaphysical question 'what makes it the case that the causal arrow points a particular way?'. We have looked at Lewis's asymmetry of overdetermination, Hausman's independence condition, Woodward's inter-ventionism, and Menzies and Price's agency. Proponents of basing cause on time will say this just confirms that we have to give up the constraint allowing backwards causation. Proponents of perspectivalism will say this just confirms that we have to give up the search for objective causation. Others will say this just confirms that we have to give up the search for a reductive account of causation. This last camp says the direction of causation is just a primitive fact in any of its instances, sui generis and irreducible (Anscombe, 1971; Maudlin, 2007). I, however, prefer to hope there is some other way past the considerations I've given. So my answer to the direction of causation, what makes it the case that C causes E rather than E causes C, is: I don't yet know.

4

On the Causal Nature of Time

Victor Gijsbers

4.1 Introduction: Time and Causation

What is the relation between time and causation? That there *is* a relation seems extremely likely, since, to give just one example, causes tend to happen before their effects rather than after them.[1] But this observation immediately raises an important question. Suppose that it is true that event *a* is a cause of event *b*; and that it is also true that *a* is earlier than *b*. What is the relation between these two facts? Can one of them perhaps be explained by the other?

David Hume, whose ideas continue to be a major influence on the philosophy of causation, answered this question in the affirmative. He gives the following definition of 'cause':

> An object precedent and contiguous to another, and where all the objects resembling the former are placed in like relations of precedency and contiguity to those objects that resemble the latter. (Hume, 1739/1978, sec. 1.3.14)

For Hume, then, causation is nothing over and above one type of event always being followed by another type of event. By definition, the cause is precedent to (earlier than) its effect. If events of type *A* are always followed in time by events of type *B*, then by definition any particular event *a* is the cause of the event *b* that follows it. The temporal relation between events is thus more fundamental than the causal relation, since causation is defined in terms of – among other things – time.

If we look at Hume's overall philosophy, it is clear why he wants and even needs to define causal relations in terms of temporal relations. Hume's crucial presupposition is that all our ideas must be based on perception, since any

---

[1] One could also point at the important role of time lapses in the perception of causation, as described in the seminal psychological studies of Albert Michotte (1946); or at the fact that philosophical discussions about the possibility of time travel invariably lead to issues of backwards causation and closed causal loops (Smith, 2018, section 3).

concept that cannot be traced back to perceptions is merely an empty word. Hume also believes that we can perceive temporal relations, but that we have no (further) perception of causation. I can *see* that I put my hand in the fire just before I felt the pain, but there is nothing additional in my experience that corresponds with a causal relation between the two events. We cannot perceive causation. But then how is it possible, given Hume's presupposition, that we have the idea of causation? Hume's answer is that causation can be defined in terms of time.[2]

Now, for Hume's strategy to work, his definition of causation must be *reductive* – that is, it must define the concept of causation in terms that are not themselves causal. For instance, if it turns out that the term "precedent" is to be defined in terms of cause and effect, then Hume would have defined causation in terms of causation and he would not have achieved his philosophical aim. If we call *causally innocent* all concepts that are not based on the concept of causation, and all other concepts *causally culpable*, then it is crucial for Hume's project that temporal concepts such as precedence are causally innocent. Let us call this claim the *causal innocence thesis*.[3]

Contemporary philosophers tend not to share Hume's conviction that all our ideas are based on perceptions. But the form of his theory of causation, namely, giving a reductive definition of causation in terms that include temporal concepts, is still the form that contemporary theories of causation generally take. Almost all the major approaches to causation – including regularity theories, probabilistic theories, causal process theories, and counterfactual theories[4] – aim to tell us what causation is by defining it in non-causal terms. The general strategy of these approaches is to provide rules for deciding which events cause which other events, given some rich enough description of the world. And for all of them, temporal relations are an important part of this description.

Again, Hume's theory can be used to illustrate this. Suppose that I'm wondering whether putting one's hand in the fire causes the pain that follows it. Hume's theory allows me to answer this question, given that I am provided with a sufficiently rich description of the world. What I need to be able to do is consider every holding-one's-hand-in-the-fire event, and then check for each of them whether they are followed by a temporally and spatially contiguous pain

---

[2] Hume's total story about causation is more complicated than this brief exposition suggests, but we need not go deeper into it for present purposes. See Morris and Brown (2017, section 5 and 6), for an overview.

[3] We cannot go deeply into methodological issues here, nor adequately discuss the relation between metaphysics and conceptual analysis. I will assume that we have temporal and causal concepts that adequately latch on to something real – time and causation. The causal innocence thesis then is to be understood as *both* the claim that temporal concepts can be possessed without the possession of causal concepts *and* the claim that time does not ontologically depend on causation. Hume certainly endorses both these propositions.

[4] See Beebee et al. (2009) for a good set of introductory essays.

event. If all of them are, if holding one's hand in the fire is always followed by pain, then I can conclude that there is a causal relation between them. (This is simply Hume's definition of cause.) Otherwise, there is no causal relation. Since during this process I need to determine the precedency and contiguity of events, the sufficiently rich description of the world has to contain temporal information.

Hume's theory wears its reliance on temporal information on its sleeve, while many recent theories do not. But the reliance is nonetheless there. For instance, it might initially seem that the counterfactual theory of Lewis (1973a) does not need spatial and temporal information, since it defines causation in terms of what would have happened in different circumstances. (Roughly, Lewis claims that *a* causes *b* if and only if *b* would not have happened if *a* had not happened.) But if we look at how Lewis evaluates claims about what would have happened in different circumstances, it turns out that temporal information is indeed crucial.[5]

So most contemporary theorists of causation are in the same boat as Hume: they can only achieve their philosophical aims if the causal innocence thesis is true. Otherwise, they are only providing us with ways of making causal judgments after we have already made them, which is of decidedly limited utility. Contemporary philosophical thinking about causation thus gives a high degree of urgency to the question whether temporal concepts are indeed causally innocent, or whether they are somehow based on or inextricably entwined with causal concepts.[6]

Unfortunately, this question is rarely asked. One may wonder why; and perhaps the answer is that philosophers, under the influence of a neo-Humean orthodoxy, have simply taken the causal innocence thesis for granted. If this is indeed the case, then the current chapter, which argues against the causal innocence thesis, may awaken us from our dogmatic slumber – or may at least ensure that our sleep is haunted by strange Kantian nightmares.

Now, there are two distinct ways in which temporal concepts could be causally culpable: temporal concepts could depend on causal concepts; or temporal and causal concepts could be mutually dependent on each other. These two options correspond to two important historical precedents: *causal theory of time*, which attempt to analyse time in terms of causation; and the argument developed by Kant, especially in his so-called 'second analogy of experience', to the effect that conceiving of the world as temporal and conceiving of it as causal are two sides of the same coin. I will argue that

[5] Lewis asks us to judge the closeness of a possible world in part by assessing the size of the spatiotemporal region where this world exactly matches ours.
[6] See Gijsbers (n.d.) for a more thorough analysis of the ways in which contemporary theories of causation require us to answer questions of causal innocence.

the second, Kantian, option – which we may call the *mutual dependence thesis* – is to be preferred.

The chapter is organised around these two historical precedents. In Section 4.2, we will discuss causal theories of time. An early example of such a theory is found in the work of Leibniz (see Futch (2008)), but they were especially popular around the middle of the twentieth century and then disappeared again in the 1970s. I will show that the arguments brought against them at the time did not warrant this disappearance. But I then go on to develop a new argument against causal theories of time: the *argument from synchronicity*. Somewhat surprisingly, a slightly adapted version of the argument also works against many forms of the causal innocence thesis.

In Section 4.3, we will go on to look at Kant's second analogy of experience. Kant argues that we cannot think of objects as placed in time unless we also think of them as falling under universal causal laws. It is a solid argument, but its reliance on Kant's philosophy of transcendental idealism makes it less suitable for convincing philosophers (and scientists) who are not committed to this metaphysical position. (And that would be nearly everyone.) For that reason, I will develop an argument of my own that is inspired by Kant, but approaches the topic from a less controversial basis. I show that insisting on the causal innocence thesis leads to an extreme form of scepticism about temporal claims, a form of scepticism that nobody wants to embrace.

If there is an a priori link between time and causation, but this is neither because temporal concepts are based on causal concepts, nor because causal concepts are based on temporal ones, then there are only two possibilities left: either there is some third concept, X, on which both causal and temporal concepts are based, or the mutual dependence thesis is true. But there are no plausible candidates for X.[7] We are thus led towards embracing the mutual dependence thesis, which should lead to a radical rethinking of our philosophical approaches to both causation and time.

4.2 Causal Theory of Time

4.2.1 Introduction

The most ambitious attempt to prove the causal culpability of temporal concepts, and the one that would deliver the most dramatic result if it were

---

[7] There *are* attempts to explain the *direction* of causation and the *direction* of time in terms of some third X, say, the direction of increasing entropy. But these attempts presuppose the concepts of time and causation. Note, for instance, that 'the direction of increasing entropy' means 'the direction *of time* in which entropy increases'.

to be carried out successfully, is the construction of a *causal theory of time*. A causal theory of time is a theory that reduces our temporal concepts to causal concepts. Here is a toy example:

> *a* is earlier than *b* if and only if *a* causes *b*.

This theory is false, since events often precede other events without being their cause: for instance, the death of Moctezuma II happened about six months before the excommunication of Martin Luther, but there is no causal relation between them. However, at least earlier events *can* causally influence later events, while the opposite is often held to be impossible. A historian *might* convince us that the death of Moctezuma did cause Luther's excommunication, through some extraordinary chain of events, but no historian could convince us that Luther's excommunication caused Moctezuma's death six months earlier. So a more defensible causal theory of time might be the following:

> *a* is earlier than *b* if and only if *a* could causally influence *b*.

Developing and defending this theory would not be easy, in particular because we would have to show that the idea of possible causal influence can be understood without a prior understanding of temporal relations. But the idea that a theory like this might be correct was quite popular among scientific-minded philosophers of time around the middle of the twentieth century, especially because of the way it fits Einstein's special theory of relativity.

In special relativity, whether one event *a* is earlier or later than another event *b* often depends on the velocity of the observer. But not always: some events are in the *absolute future* (or the *absolute past*) of *a*. For such events, all possible observers agree that they happen later (earlier) than *a*. The theory identifies the absolute future of *a* with those parts of space-time that could be reached by a signal from *a*; in other words, with all parts of space-time that could be causally influenced by *a*. The absolute past, by contrast, is identified with all parts of space-time that could causally influence *a*. So the idea that the structure of causation determines the structure of space and time is, if not quite inescapable, at least one that naturally arises within the confines of special relativity.

And thus it was that although the idea of a causal theory of time can already be found in the work of Leibniz,[8] it rose to modern prominence in the

---

[8] Van Fraassen 1970 chapter II.3 traces the history of causal theories of time through the writings of Leibniz, Kant, and Lechalas, while Mehlberg (1980) presents a more thorough treatment of the same topic. Futch (2008), chapter 5, gives a full discussion of Leibniz's causal theory of time. We will come to Kant in §4.3.2, but can remark here that his aim is not to construct a causal theory of time in the strict sense: he does not believe that we can reduce time to causation.

work of Hans Reichenbach and Adolf Grünbaum, who were deeply involved
in philosophical reflection on Einstein's theories of relativity. Reichenbach
developed two versions of his causal theory of time (1928; 1956), and so
did Grünbaum (1963, 1968). In each case, the latter version was supposed
to fix some of the problems in the former (van Fraassen (1970), chapter 6).
Van Fraassen (1970) then developed Grünbaum's theory yet further. It is van
Fraassen's version that I want to look into, since it is both relatively simple and
a representative example of the movement as a whole.

The plan is as follows. First (§4.2.2) we discuss van Fraassen's theory. We
will see that he and his predecessors make their own job very hard by limiting
themselves to a symmetric concept of causation. If we take an asymmetric
concept of causation, we can formulate a much simpler causal theory of time
(§4.2.3). We then (§4.2.4) delve into four arguments that were developed
against causal theories of time: the argument from the relative unclarity of
causation; the argument from modality; the argument from general relativity;
and the argument from causally isolated events. The first three all depend upon
a scientistic conception of metaphysics. The fourth forces the causal theorist
to embrace the Eleatic Principle, according to which causation is a criterion
of reality – but she might be completely happy to do so. Finally, I develop
a new argument against the causal theories of time, an argument based on
synchronicity (§4.2.5). We will see that this argument suggests that we cannot
reduce temporal concepts to causal concepts, but that temporal concepts and
causal concepts depend upon each other.

4.2.2 Van Fraassen's Causal Theory of Time

Van Fraassen (1970, section 6.4) bases his causal theory of time on the idea
of 'genidentity', where two events are genidentical just in case they involve
the same object. The events of me being born, me writing this article and me
dying are thus genidentical, since they all involve the same object: me. For any
object O, the set of all such events is called 'the world line of O', which we
can abbreviate as W_O.

As yet, W_O is an unordered set: there are some events, all pertaining to
the same object, but we don't know which events come before others. If we
want to make temporal claims and say that one event was earlier or later than
another, we need to introduce an ordering on W_O. Now, it may seem as if
introducing an ordering is the simplest thing in the world: we just use the
asymmetrical relation of causation. My state right now is certainly one of the
causes of my state in the near future, while my state in the near future is not

a cause of my state right now. So the ordering of the events in W_O is just the causal ordering, and this perfectly correlates with the temporal ordering of the events.

This would indeed be both possible and simple, but, for reasons that we will discuss later, Van Fraassen – like Grünbaum and the later Reichenbach – does not want to make use of an *asymmetrical* notion of causation. So he has to create an ordering in some other way.

He starts by defining the notion of a 'continuous part' of a world line. If we have a criterion for continuous parts, we can then say that b is between a and c just in case b lies on every continuous part of the world line that also contains a and c. To see this, just think about the real numbers. A number is between 3 and 5 just in case it is part of every continuous interval that includes 3 and 5. (Van Fraassen, like Grünbaum, actually introduces a more complicated scheme in order to be able to deal with circular time structures that are possible in general relativity, but we will ignore this.)

In order to define the notion of a continuous part, Van Fraassen uses the *symmetrical* concept of 'causal connectibility', as well as its opposite, 'topological simultaneity'. Two events are topologically simultaneous just in case they are not causally connectible. (In special relativity, this would mean that they are not in each other's past or future light cones, and thus that they will be simultaneous for some possible observers – hence the term.) Now, take an object O and some event e that is not part of W_O. Then, van Fraassen tells us, the set of events that lie on W_O and that are topologically simultaneous with e form a continuous part of W_O. The underlying intuition is that W_O has three parts: one that lies in the absolute past of e, one that lies in its absolute future, and then one in the middle that is topologically simultaneous with e. This van Fraassen defines as continuous, and we can see why, given that there are supposed to be no causal gaps in world lines. With this criterion for being a continuous part, and some further logical and mathematical apparatus, van Fraassen is capable of linking the events in W_O to the real numbers and proclaiming this link to be a time-coordinate assignment of the kind that physicists know and love.

The ingenuity of the theory is impressive; but it is a curious that almost all of it is needed *only* because van Fraassen starts from a *symmetrical* conception of causation. So why did Reichenbach (in the second version of his theory), Grünbaum, and van Fraassen all choose to base their causal theories of time on such a symmetric conception? We can start to understand this from the criticism levelled at Reichenbach's original 1928 formulation of the theory. Van Fraassen's objection is typical:

> The major criticisms of this theory center on Reichenbach's use of the notion of
> *cause*. Since Hume, no philosopher can afford an uncritical use of this notion.
> But even if one takes the view that the notion of causal connection is prephilo-
> sophical and that the question is not whether there are causal connections but
> how they are correctly described, Reichenbach faces a problem. For he relies
> explicitly on the asymmetry of such connections, on the distinction between
> cause and effect. If he wishes to say, like Leibniz, that by definition the 'earlier'
> one of a causally connected pair is the cause, then he must provide a criterion
> for distinguishing the cause from the effect. (Van Fraassen, 1970, 174)

To this, the appropriate response would seem to be puzzlement. If one takes
the causal relation as basic and undefinable, why not take its asymmetry as
basic and undefinable too? Historically, the answer must lie in the empiricist
commitments that Reichenbach, Grünbaum, and van Fraassen all share with
Hume, as well as their idea that the difference between a cause and an effect
cannot be observed. But one need not be an empiricist; and even if one is,
one need not hold that we cannot experientially distinguish between causes
and effects. To adapt an example from Zwart (1967, 25), suppose that I am
frightened by an evil clown who suddenly jumps from the bushes. Surely, it
is not the case that I experience both the clown and the fright but am left to
speculate about the existence and nature of causal relations between them. I
experience *being frightened by* the clown, and this experience leaves no doubt
as to which is cause and which is effect. Asymmetry is part of the common
sense concept of causation; and there seems little reason not to use it in a
causal theory of time. Doing so allows us to simplify things considerably.

4.2.3 A Simple Causal Theory of Time

Think of all events in the universe. Now order them along a line in such a
way that causes always appear to the left of their effects. Given that there are
no causal loops, this is possible. (There will in fact be infinitely many such
arrangements – simply choose one.) Imagine that the line is labelled with the
real numbers, just like the x-axis of a Cartesian coordinate system. Then we
can simply look where each event falls on the line and assign the corresponding
number to it. In this way, every event will get a number, and indeed a number
that is less than the number given to any of its effects. We can call the line *time*
and the numbers *temporal coordinates*. And that is all that time is: a numbering
system set up to reflect the causal structure of the world. This, in a nutshell, is
our simple causal theory of time.

Simple Causal Theory of Time: Temporal coordinates are nothing more
than a pragmatically useful assignment of real numbers to events in such a

way that causes are always assigned lower numbers than their effects. This is all there is to say about the nature time.

We could leave it at that, but it may be useful to work out the idea in a little more technical detail. Anyone uninterested in such technicalities is invited to skip to the next section.

First of all, we must choose a specific causal relation as the basis of our theory. For instance, we could start from the relation of *direct cause*. Suppose that I order you to throw a stone at a window; you do so; and the window breaks. Then my order is a direct cause of your throwing; your throwing is a direct cause of the breaking; but my order is an indirect rather than a direct cause of the breaking, since your throwing lies 'in between'. If we further assume that there are no causal loops, we can model the structure induced on the set of events by the relation of direct causation as a *directed acyclical graph*. This is the standard approach in the causal modelling literature (see, e.g., Pearl (2000)).

But let us instead start from the relation of *causal ancestry*, where a is a causal ancestor of b just in case there is a chain of direct causation from a to b. My order to you is a causal ancestor of the breaking of the window, since it directly causes your throwing and your throwing directly causes the breaking.[9] We will assume that the relation of causal ancestry is irreflexive (nothing is a causal ancestor of itself), asymmetric (if a is a causal ancestor of b, then b is not a causal ancestor of a), and transitive (if a is a causal ancestor of b and b is a causal ancestor of c, then a is a causal ancestor of c). In other words, the causal ancestry relation is a *strict partial order*.

Now let $t(x)$ be a function from events to real numbers such that if a is a causal ancestor of b, then $t(a) < t(b)$. That there is such a function t is guaranteed by the fact that the causal ancestry relation is a strict partial order, and by the further assumption that the cardinality of events in the universe is not greater than the cardinality of the real numbers. Indeed, infinitely many such assignments are possible. All of these are *possible time-coordinate assignments*.

We can just choose one of these assignments and call it *time*. In practice, however, some assignments are much more useful than others. So let us designate a certain class of periodic processes as 'standard clocks'. Choose an

---

[9] The advantage of the causal ancestry relation is that whether it holds does not depend on how we model the causal sequence. If we decide to add the event of the stone flying towards the window to our model, your throwing of the rock can no longer be said to directly cause the shattering; it now does so indirectly by directly causing the flying. But the relation of causal ancestry is independent of such choices.

arbitrary number Δt and impose the restraint on time-coordinate assignments that when they assign the number t to a standard clock event, they should then assign $t + \Delta t$ to the equivalent event of that standard clock's next period. (Current science has chosen the transition between the two hyperfine ground states of the caesium 133 atom as the standard clock and has set Δt to 1/9192631770.) Then let us choose one of the remaining possible time-coordinate assignments and dub it the *actual time-coordinate assignment*. We can identify time with this particular assignment. And that is all that time is: a way of assigning numbers to the events in the universe that is informed by their causal structure.

4.2.4 Arguments Against Causal Theories of Time

The idea that the temporal order simply is the causal order, and that we can introduce time-coordinates by setting up a correspondence between the natural numbers and the partial strict ordering induced on events by the relation of causation, has much to recommend itself. It seems eminently possible and it promises the kind of reductive analysis that philosophers (especially analytic ones) tend to love. So why did causal theories of time disappear from the scene? To understand this – and to decide whether this disappearance is to be applauded or decried – we need to consider the criticisms levelled against the causal theories. Three critical papers that together appear to have spelled the causal theory's doom are Lacey (1968), Smart (1969), and Earman (1972). Lacey's paper is very dependent on the details of Grünbaum's theory, and hence is of limited use to us. But it will be helpful to study the ideas of Smart and Earman.

Let us begin with Smart. He levels three main criticisms. The first is that temporal concepts are clearer than causal concepts, so that it makes little sense to try to elucidate the former by the latter. Carl Hoefer, in his very brief retrospective comments about the causal theories of time, echoes this sentiment (Hoefer, 2009, 697, n. 6), as does Earman (1972, 83). But this surely begs the question against the causal theorist. If causal concepts are unclear, and temporal concepts are causally culpable, then any simplicity that seems to attach to temporal concepts is illusory. In addition, anyone who has studied the philosophy of time will be surprised to hear that temporal concepts are supposed to be clear, rather than, say, among the most mysterious there are.

I suspect that this idea that temporal concepts are clear, while causal concepts are unclear, is based on the idea that a concept is bound to be clear if it has a well-defined use in an established scientific theory. Within the mathematical theories of physics, time is clearly defined. Causation, on the other hand, does not feature in such theories. But of course this criterion for clarity works only if one is willing to hold that the physicist's story about time captures all of time's essential features. But the deepest controversies in the philosophy of time, such as that between A-theorists and B-theorists,[10] hinge precisely on whether this claim is true.

Smart's second criticism is that causal theories of time have to use modal concepts, that is, concepts that involve possibility and necessity. *Causal connectibility* – a concept used by both Grünbaum and van Fraassen – is a modal concept since it describes the world not in terms of what is actually the case, but in terms of what *could have been* the case. In fact, Smart goes on, even the concept of causal connection itself might well be disguisedly modal. After all, we are inclined to say that if the cause hadn't happened, the effect *would not* have happened either.

Let us grant this. Why would the use of modal concepts be a problem for causal theories of time? Smart tells us that "a physical theory should be based on a purely extensional language, and the predicates '... is necessary' and '... is necessary for ...' should not occur in it" (Smart, 1969, 389). This argument seems puzzling. Surely the causal theory of time is not a physical theory, but a philosophical theory. So how could Smart's strictures on physical theories be relevant to it? The answer is that Smart is thinking of the causal theory not so much a philosophical theory about time, but as a philosophical analysis of temporal terms used in scientific theories. Again, what seems to be at work here is the idea that the essence of time is captured by physics.

Before moving on to Smart's third – and most interesting – criticism, it will be useful to discuss another argument, one that Smart mentions only briefly (Smart, 1969, 391–392) but that is central to Earman's critique of the causal theories of time: the argument, namely, that such theories cannot be made to fit the theory of general relativity. Earman's discussion is quite technical, but we can illustrate it with the following example. General relativity allows for universes in which time loops back on itself, so that by going far enough into the future (or the past), we will come back to the present. In such a universe, all events are causally connectible. Thus, constructions like the one given by van Fraassen will no longer work; and the simple causal theory presented in the previous section will also run into trouble, since the causal relation no longer induces a strict partial order.

---

[10] Roughly, A-theorists hold that the present is importantly different from the past and the future, while B-theorists hold that all moments of time have the same metaphysical status.

It might well be possible to save the causal theories of time from such problems. Van Fraassen (1972) claims so; and I suspect that the simple theory too could be made to work in cyclical time. (This would involve using the relation of 'direct cause' and the assumption that this relation is locally non-cyclical.) But suppose that Earman is right and that there are space-times allowed by general relativity that cannot be fitted into a causal theory of time. Why would the causal theorist be worried by this? Surely, she can simply claim that these models of the general theory of relativity do not describe anything we should call time. This in no way involves rejecting the theory of general relativity as a good scientific theory; it does not even involve rejecting a realist interpretation of that theory. What it *does* involve is rejecting the idea that the general theory of relativity *defines* what time is. But the causal theorist should be happy to deny that, since she holds that her own causal theory of time defines what time is.

Again and again, we see that these arguments against the causal theory of time are based on the same conviction: that physics tells us what time is and that a philosophical theory of time should thus be nothing over and above an elaboration or clarification of the physical theory. This makes sense historically. In the 1960s and 1970s, there was still a widespread distrust of metaphysics within analytic philosophy and hence an idea that if we could talk about seemingly metaphysical topics at all, it was only by linking them to established scientific theories. It thus also makes sense that these arguments made an impression and persuaded philosophers that causal theories of time might not be worth taking up. But the distrust in metaphysics has, if not vanished, at least dwindled; and it seems clear that none of the arguments discussed so far need dissuade a contemporary thinker from taking causal theories of time seriously.

This brings us to Smart's third argument, which appears to be the only one with clear contemporary relevance. He wonders how a causal theory of time can deal with events that are neither causes nor effects of other events:

> It at least *seems* to me that I can consistently envisage a universe of purely random events spread out through space-time. The objection that only if such events were causes or effects of other events could they in fact be located ought, I think, to be dismissed as too verificationist. (Smart, 1969, 394)

Let a *closed causal set* be any set of events that is closed under the causal relation; that is, a set such that if event e is in the set, then so is every cause and effect of e. Then Smart is asking us to consider a world in which there are two or more distinct closed causal sets. (In the example he gives there would even be 2^N such sets, with N the number of events in the world.) A causal theory of time will allow us to establish temporal relations between events within one set, but it also commits us to the implausible, even absurd claim that there are no temporal relations between events in different sets. How could there be pairs of events in the world that are not temporally related? Doesn't this make a mockery of the very idea that they are part of the same world?

Indeed, that is precisely what the causal theorist must stress if she wishes to avoid the absurdity. The causal theorist must claim that it is impossible for a world to contain two or more distinct closed causal sets (ignoring the empty set). Why would that be impossible? Because a world just *is* a closed causal set. To be real, to be part of our world, just *is* to be (directly or indirectly) causally connected to us: the causal theorist should embrace what has become known in the current literature as the Eleatic Principle (e.g. Colyvan (1998); Cowling (2014)). Hence, Smart is mistaken when he thinks that he can consistently envisage a universe of events that are not causally connected. (This mistake is easily made, since one imagines oneself overlooking a world of random events without realising that this establishes oneself as a causal nexus that binds all the events together as a world. Take away the very idea of an observer. Is it still clear that one is envisaging anything at all, let alone a world?)

4.2.5 The Argument from Synchronicity

So far, causal theories of time seem to be very much defensible – none of the arguments discussed in the previous section need worry the causal theorist. In this section, however, I wish to develop an argument against causal theories of time that is at the same time an argument in favour of the mutual dependence thesis.

The argument starts from the phenomenon of *synchronicity*. Carl Jung coined this word to designate an 'acausal connecting principle' (Jung, 1973) that could explain otherwise unexplainable coincidences. Thus far, my usage of the term agrees with his. But where Jung is thinking of rare, potentially life-changing events containing symbolic meaning, I am instead thinking of the most common of everyday phenomena – of the phenomenon, namely, that identical processes happen at identical 'speeds'.

Take two identical clocks and put them next to each other on a table. Let them both start out at twelve o'clock. We will witness the minute hands move clockwise until they are back at '12', while the hour hands will meanwhile move forward even more slowly to reach '1'. For each clock, there is a perfectly good causal explanation – involving their internal cogs and gears or electrical circuits – for why these particular states follow each other in this particular order. But here is the crucial question: why do the two clocks reach one o'clock at the same time? Or, to ask the question without presupposing the notion of time: why does an observer looking at both clocks always see them in the same state?

Is there a *causal* explanation of this fact? That seems extremely unlikely. To give a causal explanation of a coincidence is to give some common cause that determined that the coincidence would take place. But for all intents and purposes, the two clocks in our example are causally isolated systems. (They are not truly isolated, and they have, for instance, been in gravitational contact. But it is hard to see how such interactions could provide an explanation for their synchronicity.) We have one causally explainable process and another causally explainable process, but there seems to be no causal explanation for the fact that these two processes run in tandem.

Wait a minute!, I hear you cry. The problem is that I am assuming that causal processes are to be defined as nothing but patterns of succession. But if I describe causal processes as successions of states over certain time intervals, then the problem immediately disappears. For instance, if the basis for my causal story about a quartz clock is only that the quartz crystal keeps alternating between two states, then there is indeed no way to explain why two quartz clocks run in tandem. But if I start by stating that the quartz crystal alternates between its two states 32,768 times per second, then it's a different matter. Then I will be perfectly capable of explaining why the two clocks run at the same speed.

Indeed. But this story is not available to those who defend a causal theory of time! For if a causal theory of time is true, then assigning 1/32,768th of a second to every alternation of the quartz crystal is merely a conventional choice for assigning certain labels to certain events. But no mere labelling convention can *explain* the fact that the two clocks keep showing the same state to the observer. (It would be like explaining the clockwise progress of the clock hands by pointing out that in that direction the positions on the clock face are labelled with progressively greater numbers.) Thus, a causal theorist must accept synchronicity as an unexplained brute fact.

Is it possible to explain synchronicity while holding on to the Humean claim that temporal concepts are causally innocent? At first sight, it may seem so. The claim that the quartz crystal alternates between its two states 32,768 times per second, does not seem to involve any *causal* concepts. But on the hypothesis of causal innocence, the problem of synchronicity rears its head in a different form. For we now observe that it is only causal processes that exhibit systematic synchronicity. Non-causal processes – for instance, the appearance of the first, second, third and so on pedestrian in a given street on a given day – do not exhibit systematic synchronicity. Why is that? If the causal innocence thesis is correct, then if there is an explanation at all, it must be that we define causal concepts in terms of synchronicity. But no existing theory of causation does so, nor does this seem an especially promising direction for future attempts.

Explaining synchronicity thus seems to require a principle linking causation and time; e.g., the principle that identical causal processes always unfold during the same time interval. Neither the causal theories of time we have been looking at so far, nor theories based on the causal innocence thesis, provide us with the right kind of link. Of course, it is possible to just put synchronicity down as an unexplainable brute fact – but isn't that too high a price to pay for holding on to a metaphysical hypothesis?

4.3 A Kantian Approach

4.3.1 Introduction

The upshot of the argument from synchronicity is that neither a reduction of the causal to the temporal nor a reduction of the temporal to the causal has succeeded, since none of the extant theories is able to provide us with resources needed to explain the phenomenon of synchronicity. But even apart from the possibility that future theories might provide such resources, this argument can only indirectly imply that temporal and causal concepts are mutually dependent. To defend the mutual dependence thesis, as I wish to do, it is better to develop a direct argument to the effect that we cannot conceive of the world as temporal without also conceiving of it as causal and vice versa. We can find an argument with that conclusion in Kant's *Critique of Pure Reason*. Although the argument runs through several sections of the work (including the transcendental aesthetic, the transcendental deduction, and the schematism chapter), its core is to be found in the section on the second analogy of experience.

I first give Kant's argument (§4.3.2) and then show that it depends essentially on an acceptance of his transcendental idealism (§4.3.3). I am not convinced that transcendental idealism is wrong, but it certainly has few adherents. This means that Kant's argument won't be able to convince many people. For that reason, I develop an epistemological variation on the argument, which shows that one must accept either the mutual dependence thesis or radical temporal scepticism (§4.3.4). Since we don't want to be sceptics, we have to embrace mutual dependence.

4.3.2 Kant's Argument in the Second Analogy

Let us use the term 'apprehension' for a conscious mental image. Then our conscious sensation consists in the temporal succession of apprehensions. But this does not mean that we are apprehending an objective time order. On the contrary, the temporal succession of apprehensions is compatible with three completely different types of situation. First, there are situations like a dream, in which my apprehensions do not correspond to an object at all. I may, to use a perhaps rather worrisome example from my own experience, dream that Immanuel Kant is showing me his Star Trek museum while wearing a Vulcan mask, but then neither Kant nor the mask is grasped as a real object. Second, there are situations like looking over a house. One's gaze may first dwell on the door, then quickly sweep up to take in the roof, so that the apprehension of the door comes before the apprehension of the roof. Here there is indeed an object, but the order of the apprehensions does not correspond to an order in that object: the door does not come 'before' the roof in any objective sense. As Kant tells us:

> In the previous example of a house my perceptions could have begun at its rooftop and ended at the ground, but could also have begun below and ended above; likewise I could have apprehended the manifold of empirical intuition from the right or from the left. In the series of these perceptions there was therefore no determinate order that made it necessary when I had to begin in the apprehension in order to combine the manifold empirically.
> (B237–238/A192–193)

Third, there is the situation where the order of the apprehensions *does* correspond to an order in the object that is apprehended. Kant's example (B237/A192) is that of a ship sailing downriver; perhaps even clearer is that of an apple falling from a tree to the ground. That the apple was first hanging in the tree, then halfway to the ground, and then on the ground, slightly the worse for wear, is not some merely subjective order of my apprehensions; it is the objective temporal order of the apple's states.

In the second analogy, Kant is concerned with understanding the difference between this third situation and the other situations. Thus, he is concerned with objective time order, that is, quite literally, with the temporal ordering of (states of) objects. But his question is not, or at least not primarily, how we can *know* that a certain succession that we apprehend is an objective succession. Rather, his question is how we can *think* an objective succession. What is it that allows us to conceive of the subjective temporal succession of apprehensions as corresponding, at least in some cases, to a temporal succession of the objects? In other words, how can we think of ourselves as being in a changing world, rather than just think of ourselves as changing?

Here is yet another way to approach Kant's admittedly abstract question. What is the difference between a subjective and an objective order? Answering this question requires us to make a distinction between our subjective mental states and the objects that are apprehended in those mental states. Kant tells us that:

> appearance, in contradistinction to the representations of apprehension, can thereby only be represented as the object that is distinct from them if it stands under a rule that distinguishes it from every other apprehension, and makes one way of combining the manifold necessary. That in the appearance which contains the condition of this necessary rule of apprehension is the object. (B236/A191)

While unpacking all of this would take us too far, the core idea is clear. An object is that which makes it necessary for our apprehensions to be combined according to a certain rule – it is what ensures that our stream of mental images cannot change in any which way (as it *can* in a dream, where we are not apprehending objects). So to conceive of objects as undergoing change, that is, to conceive of an objective temporal order of events, *is* to conceive of them as necessitating a certain temporal order of apprehensions.

> If we investigate what new characteristic is given to our representations by the *relation to an object*, and what is the dignity that they thereby receive, we find that it does nothing beyond making the combination of representations necessary in a certain way, and subjecting them to a rule; and conversely that objective significance is conferred on our representations only insofar as a certain order in their temporal relation is necessary. (B242–243/A197)

Now, the idea of a necessary temporal relation, that is, of one thing having to follow from something else, is for Kant simply the idea of *causality*; and any particular such rule is a causal law. Hence, the upshot of Kant's argument is that thinking objects as placed in time, that is, thinking an objective temporal order, *is* thinking objects under the concept of causality –, in other words, thinking the world as being subject to causal laws. A temporal world that is not also a causal world is simply inconceivable.

4.3.3 Assessment of Kant's Argument

What makes Kant's argument of particular interest for a defence of the mutual dependence thesis is that it delivers a conclusion at exactly the right level, namely, that of concepts and conceivability. If Kant is right, then we really can't think objective temporality without thinking causation. Even a causal theory of time doesn't deliver a result that is quite that strong, since it leaves open the possibility (emphasised by van Fraassen) that one could also give a temporal theory of causation, and hence doesn't absolutely prove that one cannot have temporal concepts without having causal concepts. So *if* Kant delivers, he delivers exactly what we want.

Does Kant deliver? Of course, everything depends on whether we embrace his conception of objects. One could deny that there *are* objects in the Kantian sense. This would be equivalent to denying that there is any kind of necessity in our apprehensions: we just have one conscious image and then another, with the distinction between Kant's three scenarios – the dream, the house and the ship – having no real significance. Of course, this immediately shows how extreme such a denial would be. (Though arguably there are traces of it in Hume's *Treatise*.)

Of more interest is another way of going against Kant's conception of objectivity. It is crucial to Kant's argument that objects not only make certain combinations of representations necessary, but that this fact *exhausts* our grasp of what it is to be an object.[11] If we have an independent grasp of objects, the argument of the second analogy doesn't work, since there could then be a way of thinking objects as temporal that does not involve their connection by relations of necessity. In particular, it seems that we could just define objects as spatiotemporally located entities. If we did, Kant's questions wouldn't even make sense.

This becomes clear when we look at Peter Strawson's (Strawson, 1966, section 2.III.4) reconstruction of Kant's argument. Strawson believes that Kant's main line of argumentation can be separated from his metaphysics of transcendental idealism, a core part of which is precisely the conception of objects we are discussing. So Strawson, keen on giving an interpretation of Kant's argument in terms of a straightforwardly realist conception of objects as things located in space and time, tells us that Kant argues as follows. (1) There are objective temporal relations between events. (2) The causal processes of sensation generally don't switch these around; that is, if *a* happens before *b*, then we generally sense *a* before *b*. (3) The objective temporal relation between events necessitates a certain order of sensations. (4) So, we sense a necessary temporal relation between events. Unfortunately, the step from (3) to (4) is clearly invalid. Indeed, Strawson calls this 'a *non sequitur* of numbing grossness' (Strawson, 1966, 137).

Strawson's failure to find an even mildly plausible argument in Kant's second analogy lends strong support to the idea that Kant's argument about causation and time is dependent on his philosophy of transcendental idealism and on his conception of objects in particular. (This conclusion agrees with

---

[11] '[T]he relation to an object ... *does nothing beyond* making the combination of representations necessary in a certain way' (B242–243/A197), emphasis changed.

the analysis by Allison (2004).) Now, a lot can perhaps be said in favour of transcendental idealism, but it cannot be denied that it is very much a minority position in philosophy and even less popular among scientists. So for our current purposes, Kant's argument is of limited use. However, in the next section I will develop a variation on his argument that does not rely on transcendental idealism.

4.3.4 The Argument from Scepticism

We start at a less exalted level than Kant: rather than asking how we can conceive of objects and events as temporal, we will assume that we already do so. Instead, we will ask the epistemological question of how we can *know* about the temporal order of events. Later on, we will have to return to the level of concepts.

How then can we know the temporal order of events? I remember sitting in the hospital, reading in Peter Strawson's book on Kant in preparation for writing a few sentences about it; I also remember sitting behind my desk and actually writing about the book. I also know that the reading happened before the writing. But how do I know this?

As Kant tells us, 'time cannot be perceived in itself' (B233). When we experience the world, we are not experiencing it *as* at a certain time. The experience of, say, seeing a blue square in 2022 just is the experience of seeing a blue square. It is indistinguishable from the experience of seeing a blue square at any other time. So, I cannot know based on nothing but the phenomenal contents of two remembered experiences which of them happened earlier and which later.

Hume claimed in the *Treatise* that memory is distinguished from perception by being 'fainter'. Now, if older memories were 'fainter' than newer memories, or if there were some other phenomenal difference imposed upon them, then this could perhaps be used as a clue to temporal order. But this doesn't seem to be the case. Very old memories can be painfully sharp and detailed, whereas I may have only the barest recollection of yesterday's breakfast. Of course, this is not to deny that the passage of time has an effect on our memories – I would have trouble remembering the names of all my classmates back in school, whereas I used to know them very well – but it *is* to deny that we can simply compare the contents of our memories and see which of them is earlier.

What about the fact that our memories do not need to be of instantaneous states, but can be of longer intervals? I remember hearing the famous ta-ta-ta-tam 'knock of fate' motif at the beginning of Beethoven's fifth symphony. This remembrance does not consist of four separate memories about which I can legitimately wonder how their contents were related in time. I just remember hearing the motif, as such, in its temporal extension. Memories, then, can be extended in time. Does this provide us a purchase on time order? It surely does. But only of a limited nature: at the very least such memories require continuous consciousness, which means that we could never use this mechanism to find out about the time order of things that are separated by episodes of sleep. (In practice, the time that can be covered by a single episodic memory is of course much shorter. I do not have a continuous remembrance of my entire day.)

Here is another possibility: my memories may have been timestamped by the use of external clocks. Perhaps I looked at my watch in the hospital and saw it was the 8th of May, and then again behind my desk while writing about Strawson, where I saw it was the 9th. Surely this would establish temporal order? Well – as a matter of fact, I *didn't* look at my watch, but I nevertheless know the temporal order of the events. But even if I *had* looked at my watch, this would only serve to relocate the problem. For now I would have to explain how I can know that the event of my watch showing the 8th of May was temporally before the event of my watch showing the 9th. This may seem trivial, but it is not. I will return to the issue shortly.

So how *do* I know that I read Strawson's book before writing about it? In fact, there isn't much of a mystery here: reading the book was necessary for getting the knowledge needed to write about it. The reading was a causal condition of possibility of the writing. And causes always precede their effects. So the reading must have come before the writing. My certainty about the temporal order of events is based on my prior acceptance of two things: a particular causal link between these two events and the principle that causes happen before their effects.[12]

---

[12] Psychological research into assigning times to memories supports the idea that we arrive at such assignments through complicated processes of reasoning about how events hang together. Already in a 1993 review of theories about memory and time, William Friedman (1993, 58) concluded:

[T]ime information is stored not alongside memories for events but in our more general body of knowledge about time patterns. We know a tremendous amount about the temporal structure of our lives and the physical and social environment, from the daily cycle to the major events in a lifetime, and can rapidly extract the general temporal properties of some new experience [...] This general time knowledge allows us to interpret contextual information that happens to be associated with a memory. [...O]ur ability to judge the time of events is not based on special time codes assigned to memories, an inherent temporal organization of memory, or age-of-memory codes created by the passage of time. Instead, the elementary information is the ordinary contents of memory.

While Friedman does not specifically mention causal considerations, it seems clear that the body of information he mentions will essentially involve causal knowledge.

This is not a peculiar feature of this example about memories, but true about temporal judgements in general. When we are asked to judge when a particular event took place, we always use causal information. When was this bronze knife made? Certainly after the invention of bronze, since inventing a technology is a causal condition for using it. When did this animal die? It ate plants; the plants got their carbon out of the atmosphere; and the radioactive carbon has since been spontaneously decaying and has not been replenished through any process. Given the truth of all these causal hypotheses, we can use carbon dating techniques to estimate the age of the bones. How long ago was it that I looked at my watch and saw it was the 8th of May? My watch tells me it is now the 10th of May, so assuming that its internal causal processes have been unfolding as intended by the maker, that must have been two days ago. When was the event that I am currently perceiving? Right now, since the causal processes of perception take very little time – unless the event is very far away, as in the case of celestial objects.

We could multiply examples indefinitely, but let us simply draw the conclusion that our temporal judgements are generally made with the use of causal presuppositions. The crucial presupposition is of course that causes happen before their effects. If we do not invoke that hypothesis, there is no basis left for making temporal judgements. Events could be located anywhere in time. We would have to be complete temporal sceptics.

Let us illustrate this by considering a wild hypothesis: that all the events that we label as the nineteenth century happen two thousand years in the future. The eighteenth and the twentieth century are contiguous in time – so 1718 is not 300, but only 200 years ago – while the nineteenth century simply lies two thousand years further down the line. We surely know that this hypothesis is wrong. But how do we know it? Again, as Kant said, we cannot perceive time itself. What we *can* do is tell an extremely plausible story about how events in the eighteenth century caused events in the nineteenth century and how events in the nineteenth century caused events in the twentieth century. We then use the principle that causes happen before their effects to dismiss the wild hypothesis. But without that principle, we couldn't dismiss it. A world that contains the exact same events as ours, only distributed differently through time, is empirically indistinguishable from ours. (We ignore the potential metaphysical impossibility of such a world.) The only way to avoid temporal scepticism is to accept a priori the principle that causes come before their effects.[13]

---

[13] At least in general. A mild amount of retrocausation could still be accepted without making temporal scepticism unavoidable.

By itself, the necessity of accepting this principle a priori doesn't disprove the causal innocence thesis. Hume, for instance, would immediately accept the principle, since it follows trivially from his definition of cause. But the fact that we need to accept this principle in order to stave off temporal scepticism *is* a problem for the causal innocence thesis. If causal concepts reduce to temporal concepts, then it cannot be the case that causal knowledge is a guide to temporal knowledge. It cannot be the case that before I define causation, I am in the dark about temporal relations; but then, merely by introducing a definition, I suddenly have access to knowledge about the world. So, given that temporal scepticism is false, so is the causal innocence thesis. And this is what we set out to prove.

A brief coda. What is the status of the principle that causes always come before their effects? As we have seen, temporal scepticism can only be staved off if we accept this principle a priori. Received wisdom has it that all statements that can be known a priori must be analytic. But we have rejected the causal theories of time, according to which the principle is analytic because of the definition of time; and we have also rejected the causal innocence thesis, according to which the principle is analytic because of the definition of causation. But this seems to exhaust the options for analyticity. So there is a case to be made that the principle is synthetic a priori – which would mean that Kant has, after all, the last laugh in this discussion.

4.4 Conclusion

Let us recap the argument. There are four possible positions with respect to the relation between temporal and causal concepts. There could be no relation; temporal concepts could be reduced to causal concepts; causal concepts could be reduced to temporal concepts; or these two sets of concepts could be mutually dependent. The aim of the chapter has been to defend the fourth possibility.

Two arguments were developed. First, the argument from synchronicity, which shows that the reduction of temporal to causal concepts is unsuccessful. It also narrowed the range of possible reductions in the other direction, but couldn't quite exclude them. Second, the argument from scepticism, which shows that temporal concepts are dependent on causal concepts. Together, these two arguments exclude all options except for the mutual dependence thesis. I therefore conclude that we cannot think time without thinking causation and that we cannot think causation without thinking time. Having temporal concepts and having causal concepts are two sides of the same coin.

As I explained in the introduction, contemporary theories of causation almost universally assume that we can understand time before we understand causation. If this is false, we need to rethink our approach to causation. Perhaps it is time to leave the Humean phase of analytic philosophy behind and enter the Kantian phase.

5

Causation in a Physical World

An Overview of Our Emerging Understanding

Jenann Ismael

5.1 Introduction

There has been an enormous burgeoning of interest in causation across the sciences, and details of various kinds are easy to locate. One can open up a journal in microbiology and be assailed with detailed models of the causal structure of cells and proteins. One can find textbooks on the increasing array of formal and computational tools for causal search and discovery, and take classes devoted to formal methods and techniques in causal modeling. Psychologists are unraveling new details about causal learning, and how people use causal concepts in reasoning. While details of these kinds are in abundant supply, it is not easy for the philosopher or the metaphysician to find among them an answer to the question of what causation is or how it fits into a physics-based ontology.[1]

This is a survey chapter, or primer, that tries to fill that hole by

(i) assembling the pieces of our emerging scientific understanding of causation into an account of where (and how) causation arises in the architecture of the cosmos;

(ii) looking at how this emerging scientific understanding illuminates and transforms traditional philosophical questions about causation; and

(iii) considering whether and in what sense the resulting view is pragmatist.

---

[1] One of the charges that is sometimes made against the causal modeling framework, when it is presented as a philosophical account of causation by Woodward, is that the scientific questions somehow sidestep the philosophically important questions. And Steven Sloman prefaces his lovely book *Causal Models*, with an explicit disclaimer that he will not address the metaphysical questions. See Sloman (2005), and Woodward (2017) for Woodward's attempt to address those charges.

5.2 The Back History

If one looks back at the early history of science, it can be seen as developing out of systematization and abstraction of causal thinking: the search for an understanding of the causal relations among events. That is as true of physics as it is of the other sciences. I think that it would not be a mis-description of physics to say that it was devoted, in part, to uncovering the causal skeleton of the world.[2] At first, the notion of cause retained its close connection with mechanical ideas. A cause was something that brought about its effect by a transfer of force. But by the time Newton had finished with it, causation didn't appear in the presentation of his theory at all. What he put in the place of causal relations are mathematical equations that give the rate of change of a quantity over time.[3] Forces drop out of the picture.[4] There is no *direction* or asymmetry in the determination. Later events determine earlier ones as surely as earlier ones determine later ones. In 1913, Russell published a paper – "On the Notion of Cause" (1913) – that is the starting point of almost every discussion of causation and physics. In that paper, Russell noted that the kinds of laws Newtonian Mechanics gives us are so different from causal relations as traditionally conceived that it is misleading to think of them in causal terms at all. And he argued that causation is a folk notion that has no place in mature science. The modern discussion of causation in the philosophy of science really began with Cartwright's deeply influential critique of Russell's paper. Cartwright (1983) pointed out that dynamical laws cannot play the role of causal relations in science because specifically causal information is needed to distinguish effective from ineffective strategies for bringing about ends. So, for example, it might be true as a matter of physical law (because smoking causes

---

[2] I take this phrase from Russell, whose views on causation changed between 1913 and *The Analysis of Matter* (1927): "The aim of physics, consciously or unconsciously, has always been to discover what we may call the causal skeleton of the world." For discussion of his later views, see Elizabeth Eames (1989) and Kenneth Blackwell (1989).

[3] Russell uses Newton's law of gravitation as an example: "Certain differential equations can be found, which [...] given the configuration and velocities at one instant, or the configurations at two instants, render the configuration at any other earlier or later instant theoretically calculable. That is to say, the configuration at any instant is a function of that instant and the configurations at two given instants. This statement holds throughout physics, and not only in the special case of gravitation." Russell, 1913, first page. For discussion of Newton's treatment of causation, see Janiak (2013), Janiak and Schliesser (2012), and Ducheyne (2011).

[4] Bohm made the same point a little later, in his masterful book *Quantum Theory*: "It is a curiously ironical development of history that, at the moment causal laws obtained an exact expression in the form of Newton's equations of motion, the idea of forces as causes of events became unnecessary and almost meaningless. The latter idea lost so much of its significance because much of the past and the future of the entire system are determined completely by the equations of motion of all the particles, coupled with their positions and velocities at any one instant of time." (Bohm, 1951, p. 151)

bad breath) that there is a strong positive correlation between bad breath and cancer. But it is not true that bad breath *causes* cancer and hence it is not true that treating bad breath is an effective strategy for preventing cancer. And that difference – the difference between being correlated with cancer and being a way of *bringing it about* – is not one that can be drawn by looking only at dynamical laws. If one wants to avoid getting cancer, one has to know not simply what cancer is correlated with, but what *causes* it, i.e., what brings it about. Cartwright's observation led to a lot of handwringing, wondering what causal information adds to the kind of global dynamical laws that Russell took as paradigmatic of physical laws. Philosophers played with probabilistic and counterfactual analyses, and there were a lot of unresolved questions about the metaphysics of causal relations.[5]

Since around 2000, there has been a revolution in our understanding of causal thinking stemming from allied developments in a coalition of fields from computer science to psychology. And it was really the development of the Structural Causal Modeling framework by Judea Pearl (2000) and Spirtes et al. (2000), that transformed the discussion. The framework gives us a precise formal framework for representing causal relationships that is well suited to causal search and discovery in science. It has been at the center of developments in causal learning and statistical inference. It can be used to define normative solutions to causal inference and judgment problems. It has facilitated new insights into the role of causal thinking and cognition. The influence and power of the framework are undisputed and largely due to its utility in raising concrete scientific questions that can be investigated in the laboratory.

From a philosophical point of view, this – in conjunction with developments in physics that I will describe below – has helped us resolve the puzzle brought out by the exchange between Russell and Cartwright: the disappearance of causes from the fundamental level of physical description, on the one hand, and their indispensability in practical reasoning, on the other. The way that it resolves the puzzle is by giving us an articulated understanding of where (and how) causation (or, better, causal thinking and the physical structures that support it) arises in the architecture of the cosmos. And, in the way that is characteristic of philosophical questions once they pass into the hands of the sciences, it has transformed some of the old questions, while also raising new ones.

---

[5] See Hitchcock (2016) and Menzies (2014) for surveys in the development of probabilistic, counterfactual theories. See Field (2003) for a now classic article that set the agenda for metaphysicians trying to understand causation in a physics-based ontology.

5.3 The Formalism

To highlight a few relevant features of the formalism, causal relations are treated, in the first instance, as relations between types of events.[6] Type causation relates types of events like the striking of a match of a specified kind under given circumstances, rather than the particular striking of a particular match at a particular time and place. This accords well with practice in science. A causal relation between types of events is always relativized to a particular network. Networks are defined by collections of variables. Individual properties are represented as the values of variables. The choice of which variables to include and which to leave out make a difference to whether there is a causal link between a pair of variables and what the link is. Singular causal claims (e.g., the claim that a dot on a photographic plate was caused by the impact of an electron) are derived from type causation. DAGs (directed, acyclic graphs) are used to represent the causal relations among the variables in a network. Direct causation, represented by an arrow, is the most basic causal relation in a network. A variable X_i is a direct cause of another variable X_j, relative to a variable set V, just in case there is an intervention on X_i that will change the value of X_j (or the probability distribution over the values of X_j) when all variables in V except X_i and X_j are held fixed. DAGs give us a perspicuous way of displaying causal relationships. Figure 5.1 shows a couple of examples.

You can see at a glance how intervening on one will affect other variables in the network.

From the point of view of the Structural Causal Modeling framework, causal judgments in everyday contexts rely on loose and tacit assumptions about

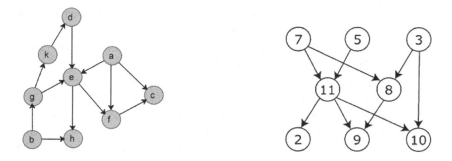

Figure 5.1 Example DAGs.

---

[6] See Pearl (2000). Also Pearl (2009) and Woodward (2003).

which variables are being considered, and what is being held fixed in situ for purposes of assessing the effects of one variable on another. Those choices are made explicit in the formalization.[7] Because it makes them explicit, it lets us separate the objective content of causal judgments from the context-dependent pragmatic factors that guide the choice of network. This means that once we have specified the choice of network and contextual factors, it is a factual question whether there is a causal relation that exists between a pair of variables, but without that specification we don't have a well-defined question of fact.

The question that comes out of the Russell/Cartwright exchange is:

How does causal information go beyond the information captured in dynamical laws, i.e., what does causal information add to those laws, or (if it doesn't add anything) what information contained in those laws does it make explicit?

Global dynamical laws give us information about relations between global states at different times. Causal information is information about the results of hypothetical interventions, where by that we mean information about what would happen if a variable is separated from its past causes and allowed to vary freely. The reason that this is not information that we can in general extract from global laws is not that global laws *don't* have modal content, i.e., it is not that they don't go beyond making claims about what is actual and have implications for what is possible. The problem is that the modal content that they have is global. Global laws tells us what kinds of world are possible. They don't tell us what would happen if we intervened in one of those worlds in a way that is contrary to law. (In logical terms, this is put by saying that the antecedents to intervention counterfactuals are nomologically impossible, i.e., they violate the physical laws.)

That gives rise to another question: how is causal knowledge possible?

If causal information is information about what would happen if some variable were separated from its causes and allowed to vary freely and no variable is ever actually separated from its causes, how could we study causal structure? The answer is that, in practical terms, there are various ways of randomizing the values of a variable, which provides an effective separation from its own value, so that we can treat changes to the values of some variables as interventions; e.g., insert a coin flip, have a random number generator set the value of an exogenous variable, or let a lab assistant choose them at will. Any variable that can be manipulated and whose own causes are uncorrelated with the variables of interest can be regarded, for such purposes, as external to the

---

[7] For these reasons, the Structural Causal Modeling framework is not an analysis of the folk concept of cause.

system, and changes in the value of such variables as being interventions.[8] This is the essence of scientific experimentation. The possibility of knowledge rests on the fact that we can effectively isolate systems in the laboratory, manipulate the external input to them, and observe the effects of our manipulations.

And this in its turns tells us something important about how we should think of (for lack of a better word) the metaphysics of causation (or better, the physical structure that support causal beliefs, i.e., the physical structures that causal beliefs carry information about).

In a world that is as complicated as ours, and which decomposes into a collection of rule-governed components, which can be isolated and studied in the lab, and where the laws governing configurations are given in terms of the laws governing components and their interactions, causal information can be extracted from laws governing components and interactions. One can concoct examples of very simple worlds, or worlds in which there are nomological restrictions on initial conditions, in which causal information cannot be extracted from *global* laws. What is really going on in those worlds is that the global dynamical laws will not allow us to recover the laws governing their components and interactions. So long as one is willing to take the local nomological structure of the world as basic, one can ground facts about the results of hypothetical interventions in that (and this follows standard practice in physics: it's a perversion of the content of physical theories to make a lunge for the global laws, and then insist that the full modal content of Newtonian Mechanics has to be somehow extractable from the form they take in our world).

5.4 Surface Structure and Underlying Processes

The relationships captured in DAGs ("interventionist causal relationships") are experimentally accessible, and often first in the order of discovery. They are part of the surface fabric of the world, and almost tailor fit to play the role that Cartwright highlighted when she argued for the indispensability of causal notions for practical reasoning.[9] But there is another tradition, which comes out of the later Russell (by way of Reichenbach and Salmon) and receives

---

[8] 'Freedom', for our purposes, can be taken to mean simply 'uncorrelated with variables of interest.'

[9] In fact, it is very natural to use that practical role as an implicit definition for causal relationships in analogy with the role that the Principal Principle plays for chance. See Lewis (1980), reprinted with added postscripts in Lewis (1986) and Price (2012a). On this understanding, causal relationships are those relationships that serve as strategic routes for bringing about ends.

its fullest development in Phil Dowe's work (see Dowe (2008)). This tradition analyzes causation in terms of causal processes. Causal processes are chains of events related by causal *interactions*, where a causal interaction is an interaction that involves the local exchange of a conserved quantity. Causal processes can be defined more directly in terms of micro-dynamical laws. Causal process and interventionist accounts are sometimes treated as competing accounts of causation, but we don't need to choose between them. They are both useful, and in classical physics we find examples of both. We have causal processes at the fundamental level, and manipulable causal relationships of the kind captured by DAGs all over the place. These are undiscriminating about levels, and neutral about underlying processes. They represent often local, scaffolded relationships among variables that can be investigated experimentally, frequently without knowledge of the underlying processes.

Both, moreover, capture some aspects of everyday notions.[10] The interventionist notion captures the idea that you have a causal connection when one thing can be used to bring another about. The causal processes capture the idea of influences that propagate through space, and they satisfy the expectation that causal connections should be mediated by mechanisms, and involve some sort of contact. If interventionist causal relationships are part of the surface fabric of the world, causal processes are part of the deep, theoretically postulated substructure that supports them.

5.5 Temporal Asymmetry

I stated that classical physics gives us both causal processes and the high-level causal networks captured in DAGs, but there is a puzzle about the relationship between the sorts of high-level causal pathways and the underlying dynamical laws that is crucial to understanding causal notions of both kinds.

Interventionist causal relationships of the kind captured in DAGs have one striking feature: they are temporally asymmetric. They run in one direction – from past to future. The temporal asymmetry is no mere matter of convention. If we follow Cartwright in saying that what is definitive of causal relationships (as distinct from mere correlations) is that they can function as strategic routes bringing about ends, it is a manifest and pervasive empirical fact about our world that all known strategic routes to bringing about ends run from past to future. Of all the reasons that Russell gave in that early paper for saying that causation had disappeared from physics at the fundamental level, this is the

---

[10] Neither captures every part of the everyday notion, which is a mix of elements drawn from causal experience, proto-theories, etc.

one that persuaded most people. The underlying dynamical laws are simply equations of motion that relate the state of the world at one time to its state at others. They have no intrinsic direction of determination.

Now, this gives rise to a straightforward physical puzzle: How do the kinds of high-level causal pathways that are an objective part of the surface fabric of our world arise from temporally symmetric underlying laws? This is not the kind of question that people in the lab studying signaling pathways in cells typically ask. Scientists who empirically investigate causal structure take it for granted that causal relations run from past to future. They set up experiments that allow localized interventions and look for the temporally downstream effects.

But it *is* a theoretical question for the physicist trying to relate the kinds of causal pathways studied in biology labs to an underlying theory. And as it happens, it has been illuminated by research into the foundations of thermodynamics. Such research began in the mid-nineteenth century and was originally focused on trying to derive the phenomenological asymmetries embodied in the second law of thermodynamics from the time-symmetric laws of classical mechanics. It only quite recently turned into a very general account of the sources of temporal asymmetry in our world. So a conversation that started about why gas will disperse to fill an open container but not collect spontaneously in one corner, and why ice melts when placed in a glass of warm water but we don't see it spontaneously form without a change in temperature, turned into a conversation about why we remember the past but not the future, and why time seems to flow from past to future, and why causation doesn't run backwards. And – as in the revolution in our thinking about causation that came with the development of the Structural Causal Modeling framework – the input from both scientists and philosophers was instrumental. There was Boltzmann and Maxwell and Gibbs (all scientists), but also Reichenbach (1956), Huw Price (1996) and David Albert (2000) (all, of course, philosophers). These are two places where the interaction between science and philosophy has been a model of productive inquiry.

There were an enormous number of false steps, and although there are many unanswered questions, David Albert's 2000 book (*Time and Chance*) brought a conceptual focus that made it possible to draw philosophical conclusions. Let me give you a very quick sketch of what Albert's account looks like.

It has three first principles:

1. the Newtonian laws of motion...
2. a statistical postulate
3. something he calls the Past Hypothesis.

The Newtonian laws of motion are the familiar time-symmetric laws that Russell was referring to. The statistical postulate is a probability distribution that assigns probability to a system's being in a given microstate, given its macrostate, that is a strict addition to the laws (it doesn't follow from them directly, though there are a lot of attempts to motivate it dynamically). The Past Hypothesis is a contingent hypothesis about the early history of the world. It says that the universe started in a state of low enough entropy to make thermodynamic generalizations applicable for the roughly 15 billion years we think these generalizations have held. The only thing you have to know about entropy, for immediate purposes, is that the statement that entropy increases but never decreases embodies all of the temporally asymmetric phenomena that fall under the scope of thermodynamic description, and maybe all of the known temporal asymmetries in our world.

The three postulates work together as follows. The Newtonian laws delimit the space of physically possible worlds. The Past Hypothesis knocks out all of those worlds that don't start in a low entropy state. If you take the initial microstates of those worlds and apply the Newtonian dynamical laws, you will see that some of the remaining worlds will be on entropy increasing trajectories, and some will be on entropy decreasing trajectories. The statistical postulate provides a probability distribution over remaining worlds that overwhelmingly favors those on an entropy-increasing trajectory. The result is that the most probable history of the universe is one wherein entropy rises.

There are a couple of reservations that one might have about this story.

Albert tells this whole story at the global level. But one need not do that. Formally, it works just as well for local adiabatically isolated subsystems of the world (systems that aren't exchanging energy with the environment), and there are reasons for thinking it might be best told at that level: (i) those are the systems to which thermodynamics is applied, (ii) it is not obvious whether the local story follows from the global one, and (iii) the probabilities have a clearer interpretation. But this issue doesn't matter here.

In addition, if you look at Albert's postulates, the Newtonian laws are time symmetric, and the statistical postulate is time independent, so the only time-asymmetric assumption is the Past Hypothesis. It's a particularly crucial part of the story because it's doing all of the explanatory work. There has been some important dispute about the right form for this hypothesis. Everyone agrees that some kind of temporal asymmetry in boundary conditions needs to be assumed to extract the temporal asymmetries of thermodynamics. In Albert's account, it is a low entropy state for the universe at some time in the distant past. Mathias

Frisch (2014), in a discussion that is specifically devoted to the asymmetries embodied in causal reasoning, has shown that there are different ways of expressing the time-asymmetric assumptions that are needed to generate the sorts of asymmetries that would support causal inferences (I think that it is right to posit that this is really the root asymmetry and that if you had it in place, the asymmetries of time and information would follow). Frisch points out that it would do just as well, for example, to assume a causal Markov condition (that any node in a DAG is conditionally independent of its non-descendants, given its parents), or an assumption of microscopic randomness in the initial, but not final state of a system. There is no substantive question as to which of these assumptions is more fundamental. They are different ways of describing the structures and the contingent asymmetries that in our world support causal inferences running from past to future. David Wallace (2012) has also argued (with a more foundational focus, but in a way that ends up making points allied to Frisch's) that in the explanatory story, what is needed to generate thermodynamic generalizations is not really the entropy but the randomness or uniformity of the early universe. So while there are still things to sort out, there is broad agreement that (pending the mathematical conjectures on which the account rests) the local macroscopic asymmetries in our part of the universe could have their origin in the state of our universe in the distant past, which developed under (effectively) classical laws into a world with the emergent macroscopic asymmetries captured in thermodynamic generalizations.

If one is interested in how this story connects to causal inference, one will be interested in extracting a Causal Markov condition.

But if one is interested in how this story connects to a notion of causation implicitly defined by Cartwright's observation that if you're given a body of correlations, the causal pathways are those that can be used as strategic routes to bringing about ends, here's how to get something that connects more naturally to that idea.[11] If we start with a thermodynamic gradient, and we hold fixed all of the information embodied in the present surveyable macroscopic state of the world (and here, I don't mean just looking around you and holding fixed the ambient temperature and dispositions of, for instance, medium-sized dry goods, I mean holding fixed all of the information in libraries and scientific databases: all of the information that we've somehow accessed and recorded, or

---

[11] This is a conception of causation that brings a connection to agency into the foreground and connects to people's pre-theoretic ideas about causation. I don't think there is a fruitful discussion to be had about what the one true conception of causation is. It's part of the general approach here to show how these variously articulated concepts arise, exhibit the practices in which they figure, and the external and internal structures that support those practices.

that we could have accessed and recorded), and we ask about the probabilistic effect of local interventions on its present state (of a kind that are possible in practical terms for agents like us), it turns out that they will propagate asymmetrically into the past and future. They will, that is to say, make a difference to the future, but leave the macroscopic past untouched. And what this means is that if we look for the strategic routes for bringing about ends – i.e. if we look for ways of doing something in the here and now that will make a difference to the probabilities of events at other times and places – we will find that there's nothing one can do to make a difference to the past, but a lot we can do to make a difference to the future (ignoring, of course, Cambridge differences such as making a liar of yourself, or redeeming early promise from later missteps...)

5.6 Causal Concepts, Causal Learning, Causal Experience

So far, everything that I have said stays at the level of physical structure. We have talked about the temporally symmetric classical laws, and the emergent asymmetric interventionist pathways that can be studied in the lab and captured in DAGs. So, we have the external infrastructure in place that describes the macroscopic fabric of the world. Now, we are in a position to talk about causal *concepts*. We've moved from talking about structure in the world to talking about something that has a role in an internal cognitive economy. To do that, we introduce an agent and we couple it to the macroscopic structure. So, we think of it as a system whose job is to extract information from the environment and use it to guide behavior, and we specify that the information it takes in is macroscopic. And we look at how causal concepts arise, which is to say that we look at the role they play in an internal inferential network that is linked to experience at one end and action at the other. The sciences that are relevant are psychology or cognitive science, and there is a wealth of fascinating work on causal cognition and learning, on the development of causal ideas, their role in inference, decision and action, as well as moral judgment.

If we step back and take a cross-section of the world, ordered by scale, and look at the layers of structure from the microscopic all the way up to the level of the human being interacting with a macroscopic environment, the picture that comes into focus appears something like this: There is the geometry of space-time, which (in a classical setting) imposes (or embodies) constraints on the causal connectibility of events. There is the matter content, the temporally symmetric microscopic laws that tell us how the state of the world at one time relates to its state at the next, and the emergent macroscopic asymmetries.

The macroscopic asymmetries support the emergence of creatures that use information to guide behavior. Eventually, they get complex enough that they do something that we would recognize as thinking. By the time we get to creatures like us, thinking is causal from stem to stern.[12]

5.7 Assessment

While there is much that is not known, and various points of dispute, the sketch that I have given here is robust with respect to most of the emerging details.[13] Notice how much it transforms both our understanding of causation and of time. It is very different from anything that might have come out of simply reflecting on the concepts. No single part of this story is the story of 'what causes are.' But I think that once all of these pieces are assembled, there is no further question that remains to be answered, no story about the metaphysics of causation that remains to be told.

There are two questions that philosophers always want to ask about causes:

(i) Where does the asymmetry 'come from'?
(ii) Are causes inside or outside the head; i.e., are they part of the mind-independent fabric of reality, or in the eyes of the observer?

We've just seen a complex answer to the source of the asymmetry. What we observed was that there is not a single asymmetry, but rather a couple of related asymmetries. The thermodynamic gradient created by a low-entropy state of the early universe creates a *macroscopic* asymmetry that supports the emergence of information gathering and utilizing systems. When such systems get sufficiently complex, information gathering and utilization takes the form of conceptual activity. In this way, creatures develop causal concepts with a built-in asymmetry that reflects an asymmetry in their practical and epistemic relationships to the world. So while there is no fundamental asymmetric relation of determination in physics,[14] for creatures like us, who couple to the macroscopic structure of the world, the fact that information flows from the past, and causal influence runs into the future is a fundamental and formative

---

[12] What I have described is the part of the stratigraphic hierarchy that is relatively well understood in physical terms. If we look below the Planck scale, the theory that we look to is quantum mechanics, and things get confusing and too unsettled to say anything definite.

[13] See Frisch (2014). The sketch is robust with respect to many of the details; it does cede to any new scientific developments.

[14] Questions about temporal asymmetry in both classical and quantum mechanics, are subtle and there is room for contention, but not in ways that make a difference to our discussion. See Albert (2000), North (2008), and Arntzenius and Greaves (2009) on temporal symmetry in classical mechanics, and Bacciagaluppi (2006) on the quantum case.

feature of our experience, one that frames every aspect of our thoughtful engagement with the world.[15]

On the question of whether causes are inside or outside the head, the right thing to say is 'partly inside, partly outside.' Our pre-scientific ideas about causation aren't articulated enough to single out one part of layered structure that leads from microscopic processes up to the coupled interaction between agent and environment in which causal thinking has its home as 'where the causes are.' Scientifically, there is no uniform usage; physicists will speak of the spatiotemporal geometry as encoding causal structure, but when biologists talk about causal structure they mean something much higher up in the stratigraphic hierarchy. When ordinary people talk about causation, they invoke ideas closely associated with the phenomenology of pushing and pulling, something with a quasi-muscular notion of compulsion in mind. We can observe how all of these ideas arise and relate to one another. I don't see that there is a factual question to be settled about which of them captures the essence of causation, and at what point the real causes come in.

Notice that this kind of account is different from what we might think of as the prototypical philosophical analysis. It's been given different names. Huw Price (2011) calls it a "genealogy," Doug Kutach (2014) calls it an "empirical analysis." It is characteristic of the way in which we get a scientific account of how these notions play a central role in mediating our practical and epistemic interaction with the world. There's an external component (the infrastructure in the environment that agents use as strategic pathways to bring about ends) and an internal component (an account of the content of causal concepts understood partly in terms of their connection to experience and action, and partly in terms of their role in an internal inferential network). A big part of fitting those two pieces together is saying how agents with the kind of information we have, and with our practical capacities for intervention, in a world that is objectively structured like ours, develop and use causal ideas.

There are two philosophically important aspects of the account: (i) it secures the objectivity of causal relations, making them apt objects of scientific study, and (ii) it's a good antidote to reifying causal relations as compulsive asymmetric relationships between localized events. There are causal pathways written into the fabric of the world, but those pathways are neither asymmetric

[15] Although the body is a physical system, and interacts microscopically with the environment, the mind is an information-processing system that receives information through perceptual channels. To say that we 'couple to the macroscopic structure of the world' I mean that the mind receives information about the current macroscopic state of the environment, and uses that information to guide interventions. From the mind's perspective, its interventions are macroscopic, because it only gets perceptual feedback on the macroscopic movements of the body, and can only *control* (in the relevant sense) what it can perceive.

nor compulsive. We see the complete account of the metaphysics of causation as given by the facts about the world and our position in it that support the acquisition and use of causal concepts. Furthermore, this account does NOT treat causal relationships as asymmetric, pushy relations of dependence that are written into the fundamental fabric of the world. But it makes causal relationships objective in all of the ways that should matter to science. It makes perfect sense to think of science as devoted to (as Russell put it in his later years) discovering the causal skeleton of the world.[16]

5.8 Conclusion

The question of how causation arises in the structure of the cosmos has been a live one since Russell observed that it appears to have disappeared from the fundamental level of physical description. In this chapter, I sketched an answer to that question that I take to be emerging from what has been learned about causation from allied developments in a coalition of fields from computer science and psychology to physics. Questions about causation span disciplines, and the workshop that produced this volume was a wonderful illustration of the need for interdisciplinary thinking in addressing them.

Acknowledgments

This chapter was originally presented at TaCits conference 2017. I'm enormously indebted the participants in the workshop, and to Samantha Kleinberg in particular, for her gracious hospitality and stewardship in bringing the volume together.

---

[16] There's a particularly interesting series of exchanges between Woodward and Price, who agree on the contours that I have sketched, on whether, and in what sense, causation is objective or 'perspectival' (Price's word for agent-dependent) on a broadly interventionist account of the content of causal claims. Woodward (2013, 2009) is concerned to protect the objectivity of causal claims, and Price (2005, 2017) points out the ways in which causal direction is relative to the agent's capacities for intervention. See also Ismael (2016).

6

Intervening in Time

Neil R. Bramley, Tobias Gerstenberg, Ralf Mayrhofer,
and David A. Lagnado

6.1 Introduction

Uncovering and describing the deep causal structure of reality is a fundamental goal of science, but it is also at the heart of human cognition. We are born into a 'blooming, buzzing confusion' (James, 1890, 462) of sensory information, and spend much of our development building up a causal model of reality that is both rich enough and accurate enough to guide us in pursuing our goals. On this view, science plays a supporting role in causal cognition: extending the domain of causal understanding beyond what can be inferred through everyday experience; integrating evidence at scales beyond the capabilities of the brain; and so helping resolve disagreements about hypothesized causal connections. However, attempts to understand the psychology of causal learning and reasoning have recently flowed in the other direction, with causal inference methods from science being used as models of causal inference in the brain. Bayesian networks have become a ubiquitous tool for modelling both non-causal and causal inference in large data sets across the sciences (Pearl, 1988). Causal Bayesian networks provide a convenient calculus for learning from, and reasoning about, *interventions* – actions or experiments that manipulate things in the world (Pearl, 2000; Woodward, 2003). As a result causal Bayesian networks have been applied to the task of modelling human active causal learning and reasoning (Bramley et al., 2015; Gopnik et al., 2004; Lagnado and Sloman, 2002; Sloman and Lagnado, 2005; Steyvers et al., 2003). Separately, a plethora of statistical methods are used for drawing causal inferences from time-series data (Friston et al., 2014; Granger, 2004), and a separate line of research explores the interplay between time, causal beliefs, and perception (e.g., Bechlivanidis and Lagnado, 2016; Buehner and McGregor, 2006).

In this chapter, we contend that the division in the scientific literature between intervention-driven and time-driven modes of causal inference is an artefact of the kinds of datasets that scientists have to work with. Experimental datasets are typically aggregated over many independent samples from a population in a setting where temporal dynamics are unavailable or unmeasured, while non-experimental datasets often track multiple factors across time. We argue that there is often no such data distinction in human experience – we experience the world as a single ongoing event stream and are constantly choosing how and when to next intervene.[1] We must be sensitive to what goes on before, during, and after our actions if we want them to be effective or informative. As such, we advocate a move toward modelling human causal learning as an inherently online process, involving the continuous integration and interplay of both temporal and interventional information (cf. Rottman et al., 2014). We illustrate limitations of treating interventions and time as separate cues, and then highlight novel approaches and data that pave the way for a new conception of the role of intervention and time in human causal learning.

The chapter is organized as follows. Section 6.2 introduces the causal Bayesian network formalism and interventional calculus, describing its successes in capturing aspects of everyday causal cognition before highlighting other aspects that it cannot capture due to its very limited representation of time. Section 6.3 introduces a new framework for representing causal structure and reasoning about interventions, where causal connections are associated with expectations about causal delays (Bramley et al., 2018b). We show how this representation helps capture several aspects of causal cognition outside the scope of Bayesian networks. Section 6.4 introduces an even richer representation of continuous causal influences between continuous valued variables, again highlighting how this can shed light on other aspects of everyday causal cognition and intervention choice. Finally, we discuss how these forms of temporal sensitivity fit into a broader picture in which structured priors, or intuitive theories (Gerstenberg and Tenenbaum, 2017), guide learning, and discuss the interplay of representations at different levels of granularity.

6.2 The Causal Bayesian Network Account of Learning from Interventions

A core challenge for causal inference both in science and in everyday human learning is distinguishing genuine causal relationships from spurious

---

[1] Although, we are also able to learn from data that takes more abstract forms, using language and other cultural artefacts to bootstrap from the accumulated knowledge of society.

or coincidental ones. Many things in the world are statistically associated – sun exposure is associated with melanin production; smoking is associated with lung cancer; ice cream sales are associated with deaths by drowning; and homoeopathic remedies are often associated with positive health outcomes. Some of these associations are due to direct causal relationships – sun exposure really causes skin to tan and smoking really causes cancer. But, in many others cases, the relationship is due to shared causal factors — people are both more likely to eat ice cream and go swimming when the weather is warm, and people often feel better after taking a medicine they believe works, regardless of whether it is actually effective (Di Blasi et al., 2001). Intervention is a way of assessing whether an association is causal. Interventions are actions that perturb the world and so reveal its structure in ways that are unavailable from mere observation (Woodward, 2003). In science we call these experiments, and plan them carefully, systematically manipulating things on a large enough scale to resolve causal questions of interest in the face of irreducible noise. When we manipulate one variable, we no longer have to worry that a resulting association between that variable and another is due to a shared cause. For instance, an experiment systematically exposing some participants to sunlight and others to darkness, measuring their skin tones before and after, will reveal a positive relationship between the intervened-on factor (sun exposure) and its putative effect (skin tone).

Interventions in human experience are more diverse and ubiquitous than those practised in science. Our every action affects our proximal world in some way, from small movements that affect our pose or field of view, to the extended actions through which we interact with other objects and one another. Sometimes we act with the goal of resolving uncertainty about a causal system ('What does this button do?', 'What does this taste like?', 'What will happen if I shout loudly in the library?'), other times we act primarily to pursue our goals (turning on the PC, feeding ourselves, getting someone's attention). Causal Bayesian networks (Pearl, 2000) provide a framework for drawing inferences based on both observations and interventions. We now briefly describe this framework and how it has been applied to the study of causal cognition.

6.2.1 The Causal Bayesian Network Framework

Causal Bayesian networks (hereafter CBNs) are parametrized graphs that capture aggregate patterns of covariation between variables in terms of a network of probabilistic causal dependencies (Pearl, 2000). Nodes represent variables (i.e., the component parts of a causal system); arrows represent causal connections; and parameters encode the combined influence of parents (the source of an arrow) on children (the arrow's target, see Figure 6.1a for an example). CBNs can represent discrete or continuous valued variables and the functional dependence between the state of an effect on the states of its causes can take arbitrary form. However, the majority of psychology research has focused on systems of binary $\{0 = \text{absent}, 1 = \text{present}\}$ variables and has commonly assumed a simple parametrization for both generative and preventative causal relationships that we describe in more detail in this chapter (Cheng, 1997).

Figure 6.1a depicts a causal network relating three binary variables X_1, X_2, and X_3, with one generative $X_1 \overset{+}{\to} X_2$ connection and one preventative $X_2 \overset{-}{\to} X_3$ connection. Under this model, X_1, X_2, and X_3 all occur with their own 'base rate' probability (i.e., due to causes exogenous to the model). However the probability that X_2 and X_3 occur also depends on the state of their causes, such that $P(X_2 = 1)$ is higher than baseline when X_1 is present while $P(X_3 = 1)$ is lower than baseline when X_2 is present.

Figure 6.1b depicts some possible observations produced by the causal system in Figure 6.1a. Any parametrized causal model s over variables $\mathbf{X} = \{X_1, \ldots, X_N\}$ assigns a probability to each datum $d = \{X_1 = x_1, \ldots, X_N = x_n\}$, meaning that Bayesian inference can be used to assess which of a set of potential CBNs (which $s \in \mathcal{S}$) does the best job of

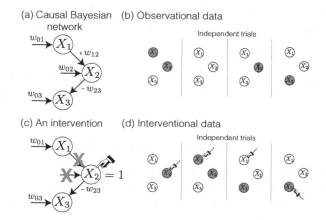

Figure 6.1 (a) A causal Bayesian network containing a generative connection ('+' symbol) and a preventative connection ('-' symbol) parametrized with base rates w_{0i} and causal strengths w_{ji}. (b) Example observational data. Shading indicates a variable was present (i.e., took the value 1) while white indicates a variable was absent (i.e., took the value 0). (c) An intervention do[$X_2 = 1$], disconnecting X_2 from X_1 and any background factors. (d) Example interventional evidence.

accounting for data. This will be whichever model best captures the pattern of statistical (in)dependencies between the variables with the minimum number of connections.

Bayesian networks embody the ambiguity described previously, about how observed correlations relate to causality. Without causal insight, we might construe the dependence between X_1 and X_2 and between X_2 and X_3 in several different ways. One is as depicted ($X_1 \overset{+}{\to} X_2 \overset{-}{\to} X_3$), but the reverse ($X_1 \overset{+}{\leftarrow} X_2 \overset{-}{\leftarrow} X_3$) is also possible, as is the case where both X_1 and X_3 depend on X_2 (i.e., $X_1 \overset{+}{\leftarrow} X_2 \overset{-}{\to} X_3$). In each case, the best-fitting parameters $\mathbf{w}_s$ differ, but the marginal and conditional probability structure is preserved and the overall goodness of fit to the data will be the same. These classes of observationally indistinguishable networks are known as 'Markov equivalent' (Pearl, 2000).

Interventions break this deadlock. An intervention in a CBN is conceived as an action that sets the values of some of the variables in the network which will then propagate through the rest of the network. This is analogous to reaching into the causal system, changing things within it, then observing what happened as a result. CBNs model interventions by fixing 'intervened on' variables to their chosen values and disconnecting them from their normal causes, using Pearl's do[.] operator (Pearl, 2000) to denote what is fixed on a given test. Figure 6.1c gives an example in which X_2 is intervened on and set to 1 (i.e., do[$X_2 = 1$]). If the model is correct, we should expect X_1 to be present with its base rate probability, while X_3 should be subject to the preventative influence of X_2 (i.e., it should be less likely to occur than if X_2 had been set to 0). Figure 6.1d gives examples of interventional evidence, under which various subsets of the variables are set through intervention either to 1 or 0 with the remaining variable states generated by running the causal system.

6.2.2 Modelling Intervention Choice

Different interventions yield different outcomes, which in turn have different probabilities under different models. This means that which interventions are valuable for identifying the true structure depends strongly on the hypothesis space and prior. For instance, fixing X_2 to 1 (do[$X_2 = 1$]) is (probabilistically) diagnostic if you are primarily unsure whether X_2 causes X_3 because $P(X_3 | \text{do}[X_2 = 1])$ differs depending whether X_2 causes X_3. However, it is not diagnostic if you are primarily unsure whether X_1 causes X_2 because X_2 will take the value 1 irrespective of its causes. Optimal experimental design theory allows us to reason about the expected value of different interventions relative to a notion of uncertainty (Fedorov, 1972; Raiffa, 1974). For instance, we can define the value of an intervention as the expected reduction in uncertainty about the true model after seeing what happened. This expectation can be calculated by averaging, prospectively, over the different possible outcomes of each potential intervention and computing the reduction in uncertainty in each case (e.g., Shannon, 1951). See Bramley et al. (2017b) for a detailed mathematical account.

Well-chosen interventions can be much more informative about a causal system than mere observations. The depicted observational data (Figure 6.1b) is much less informative than the interventional data (Figure 6.1d) with respect to the set S of all possible causal Bayesian networks relating the three variables. This is partly because the observational data does not distinguish between models with the same dependency structure, but also because a well-selected sequence of interventions can systematically target and resolve residual sources of uncertainty, for instance by testing a particular causal connection that previous tests have not provided clear evidence about.

6.2.3 Causal Bayesian Networks in Psychology

CBNs were initially adopted in psychology because of their ability to account for qualitative patterns of human judgements that are hard to capture under simple associative accounts of learning (e.g., Holyoak and Cheng, 2011; Rescorla and Wagner, 1972; Waldmann and Holyoak, 1992; Waldmann and Martignon, 1998). The idea that evidence is filtered through a causal model, provides a powerful account of human causal learning and reasoning (Cheng, 1997; Griffiths and Tenenbaum, 2005; Waldmann, 2000). Several studies have also shown that people make structure judgements from contingency information qualitatively in accordance with Bayesian network theory (Mayrhofer and Waldmann, 2016; Rothe et al., n.d.; Steyvers et al., 2009).

Pearl's interventional "do" calculus is an effective way of capturing the different ways in which people draw inferences based on observations or interventions (Sloman and Lagnado, 2005; Waldmann and Hagmayer, 2005). While observations license backward inferences, interventions do not. Observing that your colleague arrives at the office soaked suggests that it may be raining outside, while if you had intervened and poured a bucket of water on them, their wet clothes would not tell you anything about the weather. The notion that people can imagine virtual interventions helps explain important aspects of thinking. For example, virtual interventions that 'try out' different possible actions and play out their consequences are an important part of planning (Bramley, 2017; Pfeiffer and Foster, 2013). Explanation often makes reference to counterfactuals (Gerstenberg et al., 2015; Lagnado et al., 2013; Lucas and Kemp, 2015; Rips, 2010; Rips and Edwards, 2013), which are 'if statements' referring to situations contrary to fact, such as: 'If Oswald did not shoot Kennedy, then someone else did'. These kinds of inferences require a learner be able to imagine an intervention that makes the counterfactual true with minimal revision of the causal history (Gerstenberg et al., 2013; Lagnado et al., 2013; Rips, 2010).

The importance of interventional learning is well-established in education, and developmental psychology, where self-directed 'play' is seen as vital to healthy development (e.g. Bruner et al., 1976; Piaget and Valsiner, 1930). Accordingly, a number of developmental psychologists have adopted a 'child as scientist' analogy, which views children as fundamentally engaged in causal hypothesis testing within the CBN framework (Gopnik et al., 2004; Gopnik and Sobel, 2000; McCormack et al., 2016; Sobel and Kushnir, 2006). In adults, a number of studies have found that people benefit from the ability to perform (or watch others perform) interventions during causal learning (Lagnado and Sloman, 2002, 2006, 2004; Schulz, 2001; Sobel and Kushnir, 2006).

Several recent studies have also explored *how* people select what interventions to perform, comparing adults' and children's choices against the dictates of optimal intervention selection, considering constrained and heuristic variants of this (Bramley et al., 2015, 2017b; Coenen et al., 2015; McCormack et al., 2016; Steyvers et al., 2003).

6.2.4 Shortcomings of CBNs for Modelling Everyday Interventions

In this section we highlight several mismatches between CBNs and the kinds of causal inference problems that characterize everyday causal cognition. These break down into: 1) The requirement that evidence be gathered over independent trials; 2) the absence of temporal considerations relating interventions and subsequent measurements of the system; and 3) problems with representing cycles and feedback dynamics.

Independent Trials

As we have noted, CBNs are very limited in their representation of time. The interventional calculus embodies the minimal assumption that causes precede their effects, but it says nothing about the relative delays of competing causal pathways. Thus, CBNs are a natural partner to evidence that really has been gathered across multiple independent trials, in which causal relationships play out too fast to measure, or in which each variable is only measured once. By design, these features are standard in real experimental data sets. For instance, medical studies will typically involve a procedure for randomized recruitment to ensure participants are roughly independently sampled from the desired population. Then, there will be a protocol for when to measure outcome variables (such as blood pressure, insulin levels, etc.) following an intervention (i.e., delivery of a treatment). The use of a fixed trial protocol makes it straightforward to aggregate results across the sample. However, choosing an appropriate schedule of treatment and measurement seems to require substantial pre-existing causal expertise about how long the relevant mechanisms will take to work, and about when any effects will be most measurable.

A consequence of the CBN framework being widely used to study causal cognition is that many of the experiments in the literature provide learning data that is already 'packaged' into a CBN suitable format. In early studies, participants were asked to make causal judgements based on data provided in tabular form detailing how often several variables appeared together or separately (Cheng and Novick, 1990). In more recent studies, participants were shown a series of cases (Deverett and Kemp, 2012; Gopnik and Sobel, 2000; Lagnado and Sloman, 2002; Sobel et al., 2004), or invited to choose a sequence of interventions of their own devising (Bramley et al., 2015, 2017b; Coenen et al., 2015). But in these studies participants are instructed or given a cover story implying that each observation should be treated as a completely independent trial.

In some cases, the cover stories involved artificial mechanisms that reset on each trial – this includes the 'black box' toys used in developmental studies (Coenen et al., 2017b; Gopnik and Schulz, 2007; McCormack et al., 2016), but also a range of artificial mechanisms involving lights, sensors, circuits, and switches (Coenen et al., 2017b; Sobel and Kushnir, 2006; Waldmann, 2000). In other cases, the nature of the variables is not described at all (Bramley et al., 2015, 2017b; Rehder and Burnett, 2005), or independence is established by instructing participants that each trial is performed with a different sample from a population (Rehder, 2003; Rehder and Waldmann, 2017; Rottman and Keil, 2012).

While ingenious, these cover stories create situations that rarely obtain in everyday learning. The causality we encounter in physical and social systems of everyday life do not, generally, reset themselves between each interaction (Greville and Buehner, 2007; Rottman, 2016). In life, one learning episode typically bleeds into the next, with no clear boundaries. We might try a medicine on multiple occasions, but if we want to treat these tests as independent we had better leave a substantial amount of time between each test, relative to our beliefs about how long the drug takes to pass through our system. We also must keep in mind the ongoing evolution of our health and potential adaptation in our receptivity to the medication. Choosing when to act, and how to delineate between and aggregate over 'trials' in continuous experience (cf. Gallistel and Gibbon, 2000; Tulving, 1972), are important aspects of real world causal reasoning, that are missed by the studies focusing on CBN-packaged data.

Timing of Interventions and Measurements

In everyday life, observations cannot easily be separated *between* trials, and learning from actions requires paying close attention to how things play out *before, during*, and *after* interventions. Rottman and Keil (2012) point out that it is often the *change* in the state of a variable from one time point to another that people seek to explain. Furthermore, it is hard to imagine intervening on something in the world without first observing its pre-interventional state. The metaphysics of an intervention in a CBN not only presumes independence of one test from one another, but also imposes a practically implausible sequence of actions and measurements on the part of the learner. An idealized intervention involves setting variables independently from the current states of the variables in the system, essentially *before* observing their values (Pearl, 2000). Any causal effects of one's interventions are assumed to have entirely propagated through the system by the time the observation is made, while the trace of the resulting variable states must persist long enough to be measured together at the end. To make this work in psychology experiments, cover stories are used that are often either (i.) vague about the measurement of time, (ii.) pertain to variables whose states are only observable after the fact, or (iii.) involve causal relationships that would propagate too fast to be observed.

Cycles

CBNs are based on a factorization of a joint probability distribution, meaning they cannot naturally represent relationships that form loops or cycles. However, such dynamic relationships are pervasive in the world – for example an increase in the abundance of a food source (e.g., grass) causes a population increase for animals that rely on that food source (e.g., locusts) that then reduce the abundance of the food source by eating it, which then causes most of them to die out, which again allows the food source to recover, and so on (Odum, 1959; White, 2008). All sorts of real world processes, from population change (Malthus, 1888), to economic, biological, and physical interactions, are characterized by reciprocal and dynamical causal processes giving rise to emergent behaviour like periodic oscillation, self regulation, or self reinforcement. In experiments, people frequently report causal beliefs that include cyclic relationships when allowed to do so (Kim and Ahn, 2002; Nikolic and Lagnado, 2015; Rottman et al., 2014; Sloman et al., 1998; but see also White, 2008). While there are ways of adapting the CBN formalism to capture cycles (e.g. Dean and Kanazawa, 1989; Lauritzen and Richardson, 2002; and Rehder, 2017 for a recent review) these either evoke a sequence of equally spaced discrete time steps or model the equilibrium behaviour of the system. Thus, none of these proposals capture how cause–effect relationships unfold in continuous time, where some relationships might occur much faster or slower than others.

6.2.5 Summary and Discussion

In sum, as we move away from carefully organized experimental scenarios toward more naturalistic interventional learning data, the assumptions that lend the CBN framework its mathematical simplicity also make it less adequate for the challenge. When we interact with the causal world, we typically have access to a lot more evidence than independent joint state measurements. We can monitor variables continuously, tracking when events occur relative to one another and how continuous quantities ebb and flow over time. We are sensitive not just to the 'final' states of variables after causality has been and gone, but also to their states prior to interventions and their subsequent transitions. To intervene effectively and to draw sensible causal conclusions by aggregating over extended experience, we must not only worry about what variables are relevant, but also when they should be measured, how much time to leave between tests, or how to best 'reset' a system before testing it anew. This suggests that much of the important and interesting causal inference work in cognition takes place while packaging a learning problem into a CBN suitable format. Independent trials are a luxury brought about through carefully curated scenarios, and aggregation to the level of contingencies and probabilities only becomes possible when we have rich enough knowledge of functional form to abstract away from time's arrow. Thus, it is instructive to model this finer grain of human causal learning and reasoning in which time's arrow is fully represented. In the next sections, we build on these considerations and classic results about learning from temporal information, sketching two new approaches to modelling causal representation and interventional learning in continuous time.

6.3 Incorporating Time: Event Data

6.3.1 Existing Research

People have been shown to make systematic use of both event order (Bramley et al., 2014) and delay information in inferences about causal relationships (Buehner and McGregor, 2006; Buehner and May, 2003, 2004). Causal beliefs have also been shown to influence time perception (Bechlivanidis and Lagnado, 2013; Buehner and Humphreys, 2009; Haggard et al., 2002). A basic associative learning result is that, as the average interval between two events increases, the strength with which these events are associated decreases (Grice, 1948; Shanks and Dickinson, 1987; Wolfe, 1921). However, people do not always see shorter intervals as more causal, but rather prefer intervals that match their expectations, where these might come from prior experience or through familiarity with causal mechanisms. Both shorter-than-expected and longer-than-expected intervals have been shown to reduce causal strength judgements (Buehner and May, 2002, 2003, 2004; Greville and Buehner, 2010, 2016; Hagmayer and Waldmann, 2002; Schlottmann, 1999). Reliability also appears to be key to strong causal attributions from time, with variability in inter-event intervals normally associated with reduced causal judgements (Greville and Buehner, 2010; Greville et al., 2013; Lagnado and Speekenbrink, 2010, although see Young and Nguyen, 2009, for a counterexample).

Supporting the notion that temporal considerations inevitably feed into causal judgements, several studies have pitted temporal order cues against statistical contingencies. These studies generally found that causal judgements were dominated by temporal information (Burns and McCormack, 2009; Frosch et al., 2012; Lagnado and Sloman, 2004, 2006; Schlottmann, 1999). For example, Lagnado and Sloman (2006) explore a setting where a virus propagates through a network of computers infecting each with some probability, but also with variable delays in transmission from one computer to another. Participants' task was to infer the structure of these computer networks based on having observed viruses spreading through the network on multiple trials. Participants preferred causal models that matched the experienced order in which computers displayed infection, even when covariational information (which subset of computers got the virus on each trial) suggested a different structure. Furthermore, even when researchers have tried to instruct participants to ignore event timing, participants still often treated the observed timings of events to be diagnostic (McCormack et al., 2016; White, 2006).

Additionally, a result common to several of the studies that have explored causal learning through interventions on CBNs, is that participants tend to draw direct connections from intervened-on components to their indirect effects (Bramley et al., 2015, 2017b; Fernbach and Sloman, 2009; McCormack et al., 2016), ending with a kind of 'successor representation' where each variable is associated with all of its proximal and distal consequences (Dayan, 1993; Momennejad et al., 2017). This suggests that people find it unnatural for events to cause one another when they seem to occur at the same time.

6.3.2 Representing Causal Events and Interventions in Time

To account for time sensitivity in structure induction over multiple variables, we need a causal representation rich enough to encode beliefs about causal inter-event delays. Bramley et al. (2018b) recently developed a normative framework that does this. Concretely, our framework captures causality between events (hereafter 'activations') that occur at components of a system $X_1, \ldots, X_n$ at particular points in time, but have no measurable duration of their own (Cox and Isham, 1980). Bramley et al.'s (2018b) approach captures activation patterns within a causal system as being produced by a mixture of exogenous influences and endogenous causal relationships, with parametric delays governed by Gamma distributions with some mean μ and shape α. Figure 6.2a gives an example of an $X_1 \overset{+}{\to} X_2 \overset{-}{\to} X_3$ chain analogous to the CBN in Figure 6.1. As in a CBN, we can assume all three variables are activated due to exogenous factors with their own 'base rate' probability. This can be captured by a Gamma distribution where shape α is set to 1, equivalent to an Exponential distribution. This means that the expected delay before the next activation of X_1 is constant, regardless how long it has been since the last activation of X_1 or any other variable. Activations of X_1 then cause activations of X_2 with a short and somewhat variable delay (see Subfigure 6.2a, ii). These caused activations are in addition to any activations of X_2 caused by its base rate. Thus, the model implies that we should expect to see X_2 activate shortly after observing X_1 activating. Activations of X_2 then have a preventative effect on X_3. This is modelled by having activations of X_2 block the subsequent activation of X_3. Mathematically, this changes the distribution for X_3 from an Exponential back to a Gamma, doubling the time we expect to wait until seeing X_3 again but also introducing a temporary dependence between latest X_2 and the next X_3 (see Subfigure 6.2a, iii). Bramley et al. (2018b) provide the mathematical details for estimating model parameters and incorporating base rates and failure rates (similar to causal strengths in CBNs), as well as modelling conjunctive or disjunctive combined causal influences. However, here we simply highlight what we learn about causal cognition by applying this framework to human judgement and intervention patterns.

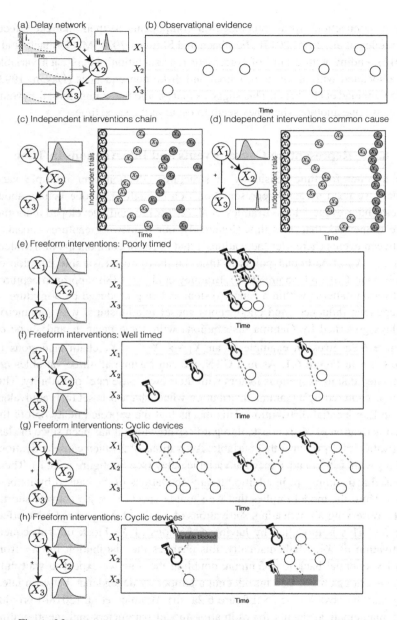

Figure 6.2 (a) Example causal delay network showing interevent delay densities for constant base rate (i.), and following a generative (ii.) or preventative (iii.) cause. (b) Rows denote variables and white circles mark activations over time. (c) Independent interventions on X_1 (rows) in a chain structure. (d) Independent interventions on a common cause structure. (e) Example freeform interventions that are not well spaced. Rows denote variables, hand symbols and thick borders denote interventions, circles denote activations. Dashed grey lines show the actual causal influences. (f) As in (e) but interventions more widely spaced. (g) Interventions in a cyclic network. (h) Including 'blocking' interventions that temporarily prevent activations of the blocked variable and so break feedback loops.

6.3.3 Modelling Inference

Unlike the CBN case, Bramley et al.'s (2018b) approach captures how timing information informs structure judgements across potentially open-ended learning periods, without the necessity of independent trials. If a learner already has a strong domain prior, for instance that a particular causal relationship will take a certain length of time to work, this will drive their judgements from temporal information in a straightforward way. That is, they will only think an event caused an outcome if the interval between cause and effect matches their expectation. However, different structures map inter-event intervals to different cause–effect delays, meaning that even without specific prior expectations about delays, the ability of candidate structures to parsimoniously account for the data can be compared, analogous to fitting a CBN to contingencies without a prior knowledge of the plausible parameters. If a learner starts out with no expectations about how long or how reliable causal delays will be, Bayesian Ockham's razor will still favour whatever structure can assign the highest likelihood to the patterns of time dependence observed between variables, while requiring the fewest separate delay parameters (Bramley et al., 2018b). For example, the reliable $X_1 - X_2$ intervals in Figure 6.2b will favour models that include a generative $X_1 \overset{+}{\to} X_2$ connection since this can explain the activations of X_2 better than a model presuming them to be exponentially distributed or to be caused by any of the other variables.

Bramley et al. (2018b) show that people spontaneously make structure judgements from event data, broadly in line with the normative framework described above. In three experiments, participants were asked to judge which of a range possible causal structures best describe the activations of components of a set of causal devices. In one experiment, participants were presented with four video clips of each causal device, each showing all of the devices' components activate over 3–4 seconds. Each video begins with the activation of the root component (X_1) followed by the activation of three other output components (X_2, X_3, and X_4) in varying orders with varying intervals. Participants not only ruled out structures that could not have produced all of the observed patterns, but also assigned more probability to structures to the extent that could have reliably produced the observed delay patterns.

While Bramley et al. (2018b) modelled the induction of general causal structure knowledge (e.g., learning that X_1 causes X_2), Stephan et al. (2018) have adopted the framework and proposed a model of causal attribution explaining how reasoners apply their knowledge about causal delay distributions to answer causal queries about singular cases (e.g., did X_1 cause X_2 on a particular occasion for which it is known that $X_1 = 1$ and $X_2 = 1$). In particular, Stephan et al. (2018) showed that causal delay information must be considered to account for the possibility of causal pre-emption of a target cause (e.g, X_1) by an alternative cause (e.g, X_3). That is, the possibility that even though X_1 was on its way to causing X_2, some other background cause X_3 might have 'gotten in first' and already caused X_2. This is a problem that singular causation judgements are highly sensitive to, but that cannot be handled by preexisting accounts of causal attribution.

6.3.4 Modelling Interventions

Just as with covariational observational data in Figure 6.1b, pure observations of events in time cannot unambiguously reveal causal structure. For example, it could be the case that the tight temporal relationship between X_1 and X_2 in Figure 6.2b comes about because the variables share an unmeasured cause that brings about X_1 faster than it brings about X_2. Fortunately, intervention again makes it possible to resolve these kinds of ambiguity. Any temporal dependence between the intervened-on variable and activations of other variables suggests the presence of a causal influence. For this to hold, interventions must be performed at random or prechosen intervals so as to be temporally independent of any alternative causes. Concretely, for two variables X_i and X_j, if the distribution of inter-event delays $p(t_{X_j, X_i})$ has a best-fitting α parameter > 1, this means that X_j hastens X_i relative to its baseline, implying that X_j (directly or indirectly) influences X_i. A key property of Bramley et al.'s (2018b) modelling framework is that it captures how the sequence of events following an intervention carries information about the underlying causal structure. Post-interventional event *order* places constraints on what caused what on a given occasion. But beyond this, the causal delays inherit time variability from their parents relative to interventions. Thus, even if an intervention causes the same activations in the same order each time, if the timing of one of these downstream events carries information about another later one, this tells us something about the causal ordering of the variables. Figures 6.2c and 6.2d visualize evidence produced by interventions on the root component of a chain and a common cause device. The device in Figure 6.2c is in fact a chain where $X_1 \overset{+}{\to} X_2 \overset{+}{\to} X_2$ with two fast connections, one unreliable connection from $X_1 \overset{+}{\to} X_2$ and another reliable connection $X_2 \overset{+}{\to} X_3$. This is revealed by the data for the chain structure: if X_2 happens late, X_3 tends to happen late as well. However below we see a $X_3 \overset{+}{\leftarrow} X_1 \overset{+}{\to} X_2$ with one unreliable fast connection $X_1 \overset{+}{\to} X_2$ and one reliable slow one $X_1 \overset{+}{\to} X_3$. Here the $X_1 \overset{+}{\to} X_2$ and the $X_1 \overset{+}{\to} X_2$ delays are uncorrelated.

We can think of the propagation of variability through the system as acting akin to a game of 'broken telephone'. In broken telephone, players form a line and then whisper a message from one person to the next, which becomes increasingly garbled along the way. Knowing what message a player heard tells you roughly where in the line of people they were. Similarly, the accumulation of errors in a signal (here in the lateness or earliness of the event timing) can reveal the causal sequence. As with broken telephone, if there is no noise (if everyone passes the message perfectly; or if all causal relationships are perfectly reliable) there is no way to tell in what order the message was passed on. But, a moderate amount of noise can lead to a 'blessing of variability' effect.[2] Bramley et al. (2018b) show that people are sensitive to this subtle signal, favouring chain structures for a range of scenarios in which a mediating variable carries information about a distal effect X_3 and common cause structures when it does not.

6.3.5 Choosing When and Where to Intervene

Bramley et al.'s (2018b) approach is applicable to situations where variables may activate or be intervened on multiple times during a single learning period, and in which the underlying structure might contain cycles. Thus, Bramley et al. (2017a) also used the framework to explore learners' choices of interventions in extended interactions with unknown causal systems. Instead of giving participants a sequence of independent trials in which to set variables as in traditional interventional learning studies, participants were given a short time to interact with a dynamic causal system and perform a small number of interventions by clicking on components on the computer screen to activate them.[3] In this setting, learners had to choose both when and where to intervene. Bramley et al. (2017a) explored learning about a mixture of acyclic structure (no feedback loops) and cyclic structures (containing at least one feedback loop), and contrasted learning about devices whose causal delays were known to be reliable (1.5 seconds with standard deviation 0.1 seconds) with those that were known to be unreliable (1.5 seconds with standard deviation 0.7 seconds). Causal connections had a small failure rate and there were no

---

[2] The same thing holds for covariational data under certain parametrizations. For example with no background noise and weak causal connections the latter links in a chain become progressively less likely to be present.

[3] See https://neilrbramley.com/experiments/it/videos/example_trial.html for a video of a trial.

exogenous activations aside from the learners' own interventions. Consistent with the predictions of the normative model sketched above, Bramley et al. (2017a) found that participants were able to identify the majority of the causal connections based on the delayed activation information they produced, and were more accurate at identifying the structure of the devices when the delays were reliable.

Bramley et al. (2017a) found that successful participants tended to leave large and regular intervals between each intervention, especially when the true causal relationships were unreliable. Successful participants also tended to wait longer after the most recent activation before intervening again, suggesting they waited until they had a low expectation that any other confounding activity would occur during their intervention. For example, Figure 6.2e shows a poorly chosen set of interventions that occur close together, making it ambiguous which activations caused which. By contrast, the well-spaced interventions in Figure 6.2f make it clear that X_1 is a common cause of X_2 and X_3. Both X_2 and X_3 reliably follow from the two interventions at X_1 but, on the second occasion, X_2 and X_3 reverse their order of activation, essentially ruling out the $X_1 \xrightarrow{+} X_2 \xrightarrow{+} X_3$ chain hypothesis that was plausible up to that point. No study has explored preventative causation in the domain of continuous time causal events. However, by contra-position to the above, the ideal time to test for a preventative relationship (as in the $X_2 \xrightarrow{-} X_3$ link in Figure 6.2a) should be when one has a strong expectation the effect is otherwise about to occur.

For the acyclic devices tested in Bramley et al. (2017a), participants exhibited a substantial preference for intervening on the root components, an effect also found in CBN-based studies (e.g., Bramley et al., 2015; Coenen et al., 2015). Unlike in most CBN-based experimental settings, these positive test interventions *are* informative about the relationships between the downstream variables due to the blessing of (moderate) variability described above. By intervening at the root of the causal device, learners can observe how the causality propagated through the system making use of ordering and delay correlations to uncover the causal sequencing. Thus, it could be that positive testing effects are largely a result of the overapplication of a strategy that works well in the real world where we can normally expect to experience the effects of interventions playing out over time.

Bramley et al.'s (2017a) experiment is also the first comparison of human learning about acyclic and cyclic devices in continuous time. Participants' interventions on cyclic devices often led to sustained patterns of activation that continued until a failed connection caused the system to acquiesce. This meant that participants experienced substantially more events in total even though they tended, reactively, to perform substantially fewer interventions on the cyclic devices. While this resulted in stronger evidence about structure for the cyclic problems according to the normative learning model, the data was also much more computationally taxing to deal with in real time given the variability in the true causal delays. The pattern in Figure 6.2g demonstrates this. It is clear from the repeated activations that X_1, X_2, and X_3 are related in some cyclic way. However, establishing the exact set of connections (i.e., whether X_2 or X_3 feeds back to X_1) is much harder. There are also many plausible patterns of actual causation that could have given rise to this data – that is, there are many ways one could draw dashed lines as in the figure, to attribute each activation uniquely to a unique previous activation or intervention. To estimate the posterior probability distribution over potential structures, the model needs to sum over large numbers of patterns of possible actual causation (Halpern, 2016). Thus, it is not surprising that human learners with limited cognitive resources, were substantially worse at identifying the structure of the cyclic devices. Bramley et al. (2017a) account for this result by positing that human learners build up their causal models incrementally, reacting to each new activation by attributing it to a recent intervention or activation. When there are many causal influences going on simultaneously, this approach becomes less accurate.

6.3.6 Summary and Discussion

We have developed a normative framework for representing causal structure relating events that unfold in time, and used it to explore real time causal learning and intervention selection. Applying this framework to experimental data demonstrated that people use temporal delays between events systematically to infer the causal structure that most parsimoniously explains what happened. The framework also showed that people are able to choose sensibly *when* to intervene, so as to not confound the consequences of their interventions with the unfolding dynamics of the system. The framework captured why people have a preference toward initiating the root component of causal systems – unlike the CBN-packaged data, this more natural temporally extended data often contains evidential signals about the structure of causality as action propagates though the system.

The analyses also shed light on the intuition that cyclic systems are naturally harder to learn and reason about. The computational cost of exact inference in Bramley et al.'s (2018b) framework increased with the number of events that might be related, and cyclic systems tended to produce more activations and non-causal regularities. The interventions investigated by Bramley et al. (2017a) can be thought of as 'shocks' to the system, where an additional event is injected into the system by the learner. One way that learners might get clear evidence about dynamic systems in real world contexts is by using extended interventions to 'block' rather than just 'shock' variables. By 'blocking' we mean holding a variable in a particular state for an extended period, while 'shocking' instantaneously changes the value of a variable but does not hold it in that position. In the current punctate event context, a blocking intervention could be to prevent a variable from activation thus disconnecting it from participating in the causal process.[4] Figure 6.2h considers a case in which a hypothetical learner uses a mixture of interventions similar to above, with extended 'block' interventions that prevent the blocked variable from activating for a period of time. This 'blocking' action is analogous to setting a variable to 0 in the CBN setting explored in 6.2. Blocking breaks the feedback in the system allowing the learner to identify one part of the structure at a time as they would in the acyclic cases. As we discuss at the end of the chapter, future work could explore how people use a combination of 'shocks' and 'blocks' to learn about cyclic systems.

Lagnado and Sloman (2004) proposed that real-time interventions exhibit a strong order cue: events that happen shortly after interventions are likely to have been caused by these interventions. Bramley et al.'s (2018b) framework captures under what conditions this heuristic works well, but also formalizes how interventions provide structure evidence regardless of whether this holds. If the base rates of variable activations are low relative to the duration of actual causal delays, interventions that activate the cause will tend to create a reliable order cue. That is, effects of the intervention will frequently be the next activations to occur as they are in Figure 6.3e. The fact that participants struggled when there was a dense flurry of activations in Bramley et al. (2017a) is suggestive that people really do rely on this kind of cue to build up their hypotheses. However, the normative framework also captures how inference is possible even when this does not hold. Wherever interventions on one variable affect the distribution of delays between that variable and another, either hastening or delaying them relative to their baseline, this is evidence for some form of causation.

Unlike with a CBN, a learner who forms a time-extensive representation is also positioned to make real-time predictions about ongoing dynamics

---

[4] In continuous variable contexts like the one discussed in the Section 6.4 'blocking' has a different interpretation: namely holding a variable at a particular value for an extended period.

(Clark, 2013). This allows them to allow for ongoing causal processes in choosing when to intervene. This was evident in Bramley et al.'s (2017a) experiment in which successful participants waited for activity to die out before intervening again.

One weakness of Bramley et al.'s (2018b) approach is that normative inference scales rapidly with the density of events experienced relative to the length of the causal delays, meaning that cyclic or multiply connected structures become very hard to evaluate. One has to reason about many potential paths of actual causation playing out simultaneously. Fortunately, as the number of potentially related events becomes unmanageable for reasoning at the level of one-to-one cause–effect mappings, one can start to reason at a slightly rougher grain, about whether activations and interventions affect the *rates* of occurrence of other variables. Pacer and Griffiths (2015) model this setting, reanalysing an experiment from Lagnado and Speekenbrink (2010) in which the occurrence of several types of 'seismic waves' could temporarily alter the rate at which earthquakes occur. This approach matches with a range of findings suggesting that beyond a small number of unique entities, people switch to approximate counting, known as 'subitising', rather than attempting to focus explicitly on each individual entity (Mandler and Shebo, 1982).

Another limitation of Bramley et al.'s (2018b) approach is its assumption that causality relates to point events. This idealization seems reasonable for processes where the causal influences are occasional and sudden – such as spiking neurons, earthquakes, gunshots, explosions, and so on. However, many other causal influences are continuous in time, and many causally relevant variables can take a continua of values with no transition or state being causally privileged. Thus, in Section 6.4 we extend the continuous time framework to capture learning and inference about variables that evolve and affect one another continuously in time.

6.4 Incorporating Time: Continuous Variables

While events are often a natural level for reasoning about the world, they need not be construed as discrete or punctate, and change need not be thought of as being constituted by *events* at all (Soo and Rottman, 2018). Many variables change continuously over time without clear change points, and their consequences can often 'linger on and mix with the effects of other actions' (Jordan and Rumelhart, 1992, 224). This means that in many contexts, better causal evidence can be had, and finer grained causal predictions can be made, if one represents the causal world as a system of continually interacting continuous-valued variables. Many real-world applications of causal inference methods focus on this kind of data. Financial forecasting, climate research, and many aspects of structure estimation in neuroscience involve reasoning about causal influences between continuous variables. The most well-established of these approaches is the so called 'Granger causality', which uses time-delayed regressions to assess whether one variable predicts the later values of another (Granger, 1969). Soo and Rottman (2014) show that such change–time asymmetries are an important determinant of human causal judgements. However, this approach suffers from the same limitations as any observational method for assessing causality.[5] Frequently variables *Granger cause* one another without truly causing one another, due to their both being influenced by some unmeasured common cause. For example, because neural activity waxes and wanes due to natural circadian rhythms, regressing one neural signal on another, even with a short delay, will often result in a significant correlation because both measurements share patterns of variation that stem from the brain's overall energy consumption rather than from specifically directed communication. A number of studies have also explored how people learn continuous valued functional relationships more generally, but not in the context of inferring causal structure (Griffiths et al., 2009; Pacer and Griffiths, 2011; Schulz et al., 2017a). So, in this last section, we survey a framework for modelling interventional causal learning about continuous valued variables that influence one another in continuous time. Note we are not arguing this framework supersedes that of Bramley et al. (2018b), but rather that it is able to capture another form of time sensitivity that is appropriate for some learning contexts but inappropriate for others.

6.4.1 Representing Continuous Causality

Davis et al. (2018a) recently developed a new framework for modelling networks of continuous valued variables interacting causally with one another. Their approach is based on the Ornstein-Uhlenbeck (OU) process (Uhlenbeck and Ornstein, 1930). OU processes describe something like Brownian motion that is subject to a corrective force that increases the further the variable strays from its mean and have been used to model a variety of phenomena including physical (Lacko, 2012) and financial (Barndorff-Nielsen and Shephard, 2001) systems as well as attention during multiple object tracking (Vul et al., 2009). Davis et al.'s (2018a) insight was to treat all the relationships in a causal

---

[5] Although, see Eichler and Didelez (2010) for preliminary work linking interventions with estimates of Granger causality.

network as simultaneously evolving OU processes, such that each variable is noisily attracted to some function of the current states of its cause(s). To model ongoing causal relationships, the mean attraction state for a variable $X_i \in \mathcal{X}$ is defined as some function of the latest value of its putative cause variable(s) X_j^t.[6] Davis et al. (2018a) focus on linear relationships, with a single multiplicative parameter w_{ji}, such that the updated value of $X_i^{t'}$ is given by

$$X_i^{t'} = X_i^t + N\left(\theta\left(\sum_{j=1}^{ca(i)} w_{ji} \cdot X_j^t - X_i^t\right), \sigma\right) \quad (6.1)$$

where σ is the noise level and θ controls how strongly the cause(s) influence the effect. Thus, if $w_{ji} > 0$, X_i will be attracted to some positive fraction of the value of the latest value of X_j if it has just one cause, where this is assumed to be the sum over all causes $j \in c(X_i)$ if there is more than one. In line with the previous sections, we call this a 'generative' connection as it implies that increases in X_j cause increases in X_i. If $w_{ji} < 0$, X_i will be attracted to some negative fraction of the latest value of X_j, with X_i tending to decrease as X_j increases and vice versa. Values of w_{ji} between -1 and 1 will make X_i undershoot the absolute value of X_j, while values greater than 1 or less than -1 will make X_i overshoot. There is no causal relationship from X_i to X_j if the attraction strength $\theta_{ji} = 0$. If a variable has no causes, its movement can be modelled as a classical random walk or as tending to revert toward some static mean value. Davis et al. (2018a) assume multiple causal influences sum in determining the attractor state for the effect. Figure 6.3a shows an example continuous variable network $X_1 \xrightarrow{+} X_2 \xrightarrow{-} X_3$ analogous to those explored in Sections 6.2 and 6.3. In it, X_1 has no causes. X_1 is then a generative cause of X_2 which is a preventative cause of X_3. The subplots visualize the distribution of possible X_i' values for each variable relative to the variable's current position and the current position of its cause.

Figure 6.3b shows example data generated by the network in Figure 6.3a. The root component X_1 (solid black) drifts around randomly, while X_2 (dotted dark grey) chases noisily behind X_1, and X_3 (dashed light grey) chases the inverse of X_2. As with previous settings, it is not possible to tell for certain what the right causal relationships are, based on this purely observational data.

---

[6] Here we express OU dynamics between small but discrete equal time steps. However, in its general form an OU process is described by a stochastic differential equation, making it truly continuous over time.

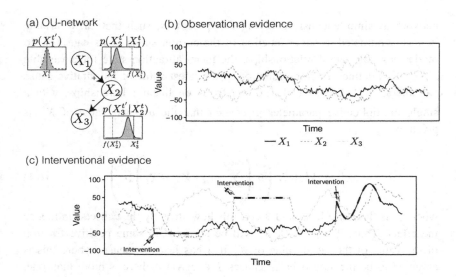

Figure 6.3 Observations and interventions on a causal chain relating continuous variables in an Ornstein-Uhlenbeck network, adapted from Davis et al. (2018a). (a) Visualisation of network and transition probabilities between t and t' relative to states latest states of effect X_j and cause X_i. (b) Observational evidence in which OU-network from a is simulated forward. Solid black line = X_1, dotted dark grey line X_2, dashed light grey line = X_3. (c) Example of interventional evidence. Heavy black dashed lines indicate interventions that hold variables in position or move them systematically.

This is partly because all three variables could be effects of some common cause with different time lags (i.e., with different θ values). Additionally though, the sustained corrective causal influences mean the downstream variables X_2 and X_3 rapidly asymptote to be close to their target values, making it hard to see which moved first unless X_1 happens to move dramatically.

One interesting property about these continuous time networks that is not captured by the event networks in Section 6.3 is their ability to produce rich emergent feedback dynamics of the kinds observed in the natural world. For example a pair of variables with bidirectional influences that undershoot one another (i.e., $w_{ji} = w_{ij} < 1$) creates an inhibitory feedback loop where both variables trend toward and then stay around zero no matter their starting point, while settings where the feedback leads to progressively larger overshooting (i.e., $w_{ji} = w_{ij} > 1$), excitatory feedback, where the variables rapidly approach positive or negative infinity. Feedback with opposite signs (i.e., $w_{ji} = -w_{ij}$) leads to oscillations such as those seen in predator–prey dynamics, where variables chase one another up and down in oscillations that can either acquiesce or can increase in magnitude over time. Since these kinds of emergent dynamics are common in the natural world, it is valuable to have a simple representation that is able to mimic them. However, it remains to be seen how commonly observation of high level dynamics uniquely reveals the internal structure of the system, as it may be that there are multiple ways of setting up a system of variables to exhibit the same emergent behaviour.

6.4.2 Modelling Inference and Interventions

Davis et al. (2018a) devised a simple experiment to explore human interventional learning about continuous valued variables in continuous time. As in the studies discussed in Sections 6.2 and 6.3, participants were tasked with identifying the structure of a number of causal devices by interacting with them on the computer screen. As in Bramley et al. (2017a) they were given a short period to interact with each device. However, rather than being visualized as clickable nodes, the variables were visualized with vertical sliders. The underlying network could contain generative $w = 1$ and preventative $w = -1$ connections and feedback loops. Unless intervened on, the sliders' positions updated continually, appearing to jitter and track up and down as in Figure 6.2b. By intervening and moving the sliders, participants could hold variables at desired values or drag them up and down.[7]

Through interacting with the systems, participants were able to identify the large majority of the connections. As in the other causal learning experiments discussed, the most frequent error was a failure to distinguish direct and indirect causal influences. Participants would often mark direct connections from intervened on variables to indirect effects. Due to the convenient mathematical properties of OU processes, Bayesian inference could again be used to estimate a posterior for each trial and used as a normative standard against which to compare participants' judgements. Davis et al. (2018a) compared full Bayesian inference against a local computations model that assumed learners built up their models by focusing on one pair of variables at a time rather than the whole structure, finding that around two thirds of participants were better described by this heuristic (cf. Fernbach and Sloman, 2009). The normative analysis also revealed that with only one exception, participants interventions dramatically increased the strength of the available evidence.

The role of interventions in this continuous context intuitively has an additional property compared to the previous contexts. Because the variables

---

[7] See https://neilrbramley.com/experiments/ctcv/demo.html to try an example trial.

are capable of taking a wide range of values but naturally move only a little per time step, interventions allow learners to inject dramatic swings and signals into the system, or hold variables at extreme or unusual levels (Figure 6.3, Davis et al., 2018a). For example, Figure 6.3c shows a learner performing several dramatic interventions. First they intervene to hold X_1 at 50 (here, for 40 'timesteps'). This results in a marked increase in X_2 and, with more noise and lag, a reduction in X_3. They then intervene to hold X_2 at -50, which only increases X_3, while X_1 returns to moving randomly, suggesting the structure is a chain. Third, they intervene on X_3 in a sinusoidal pattern that leads to a (lagged and noisy) sinusoidal pattern in X_2. While Davis et al. (2018a) did not analyse participants' precise intervention patterns, a large proportion appeared to involve moving and holding variables at extreme values, or rapidly dragging them up and down the full range of the sliders. It is worth noting that in this setting, interventions naturally act as both 'shocks' and 'blocks'. Intervening on a variable for an extended period not only takes over its state but also blocks it from participating in feedback dynamics.

6.4.3 Continuous Time Control

The interventions participants could perform in Davis et al. (2018b) were often extended in time, meaning participants were able to react to the movements of the other variables during their intervention, for example they might be manipulating one variable in such a way as to keep another (effect) variable in a particular position. This reactivity brings the interventional learning problem increasingly close to an adaptive or 'dual control' problem (Feldbaum, 1960; Guez, 2015; Klenske and Hennig, 2016; Schulz et al., 2017b). This means one faces the 'dual' problem of learning how the system works while already attempting to control it. For example, we learn to play tennis, largely, while attempting to play tennis. At first our shots are wild and do not go where we intend. But, we slowly learn to adapt our swing to different angles and speeds of the incoming ball hopefully building a causal control model of tennis in the process.

A key question therefore, is to what extent people will spontaneously learn causal structure while attempting to master the control of a causal system. To investigate this, Davis et al. (2018b) used their OU driven causal models as a class of control problems. Their task was similar to that in Davis et al. (2018a), except that participants' interventions were restricted to one 'control' variable, and limited single step increments up or down. Participants' goal was to keep another 'target' variable in a reward region. To master this task in a model-based way, a participant would have to learn the structure of the network and use this knowledge to plan the actions that maximize the time that the target variable stays in the reward region. Overall, participants learned to control in most causal systems well above chance, and learning profiles were broadly consistent with the idea that they first explored then exploited the causal model. Future work will need to carefully probe what representation participants form during a control period.

6.4.4 Summary and Discussion

In summary, networks made up of OU-related continuous variables provide a powerful extension of causal models for reasoning about fine-grained contin- uous time relationships, as well as providing an interesting test bed of control environments. The dynamics of these systems map onto real-world problems in intuitive ways and display the kinds of emergent properties we see in real world dynamic causal systems. These networks also have convenient mathematical properties that make normative inference tractable and so allow assessment of the evidential value of interventions. OU processes are unique in being Gaussian, Markovian, and stationary, all of which make likelihood estimation closed form and straightforward (Uhlenbeck and Ornstein, 1930). People were able to learn robustly through intervention in this setting, suggesting that they are not limited to reasoning about causality at the level of independent trials or even of real time event cascades.

Interventions in these systems can be used to create dramatic signals that are hard to mistake if they appear in the dynamics of downstream variables. Partic- ipants were generally sensitive to this in that they used dramatic swings or held variables at extreme values, which allows for maximally strong causal signal that is easily spotted as it propagates to other variables. However, the richness and immediacy of the evidence also seemed to push participants toward adopt- ing a local strategy focusing on pairs of variables at a time and so struggling to infer the correct structure in the case of indirect causation. As with the event networks, one way to use interventions to get clearer local information in this setting would be to allow combination interventions in which one variable is held in place while another is wiggled around. For instance holding X_2 stationary and moving X_1 up and down in the example in Figure 6.3 would make it easy to establish that X_1 does not directly influence X_3 except through X_2. Intuitively, this is a very natural kind of intervention to perform in everyday life, but one that is not easily captured by the simpler and more abstract notions of intervention and causal influence embodied by the CBN or even the delay networks in Section 6.3.[8] This highlights a key property of real-world interventions. Sometimes it is possible to insert a unique signal with an intervention by setting a variable to a highly unique value or moving it in a highly unusual way. The subsequent inferences are then a kind of message detection, where one looks for traces of the original signal reappearing in other variables.

6.5 General Discussion

There are a number of reasons why CBNs provide a good starting point for thinking about causal cognition. However, we have argued that there are also fundamental reasons why they do not provide a fully adequate account. CBNs represent time only as a partial ordering of influence, while real-world learning contexts demand a deeper sensitivity to time's arrow. We highlighted recent work that introduces new formalisms and uses them to explore the ways in which human causal learners are sensitive to time. This revealed sophisticated time sensitivity in inferring causal structure but also in timing interventions and control. The representation developed by Bramley et al. (2018b) captured causality relating events in time and showed how a preference for reliably timed causal influences can guide structure inference, as well as how sensitivity to the possibility of delayed effects shaped participants' intervention behaviour. The extended representation, developed by Davis et al. (2018a), modelled causality at a finer grain, in terms of ongoing influences between continuous-valued variables. Comparison of human learners in such a setting revealed that people are capable of interpreting rich continuous causal evidence, and are adept at using the full range of relevant variables to generate distinctive interventions whose signatures could be tracked as they propagated through the system.

6.5.1 Interventions in Rich Domains

Much of the work on intervention selection has used the metaphor of optimal experimental design (Fedorov, 1972). However, exploring richer interventions and learning of richer causal representations seems to suggest a different metaphor (Coenen et al., 2017a). We can think of interventions in richer temporally extended settings as injecting signals into causal systems, similar to how a plumber might pour dye down a sink while trying to figure out a network

---

[8] In a CBN, we could equate this to performing a set of independent multi-hold interventions where X_1 is varied while X_2 is always fixed to the same value (Woodward, 2003).

of pipes in an old house. Wherever the dye shows up must be downstream of the sink, and the length of time it takes to get there says something about the network of pipes in between. When the propagating signal is not very distinct – such as the time of occurrence of an event as in Section 6.3, or the final states of the variables after a causality has been and gone in Section 6.1 – collecting multiple approximately independent trials is still important. However, the learner again must be able to use her knowledge about the domain to create situations that are approximately independent and identically distributed.

One other consideration is that punctate interventions, or 'shocks' allow a dynamic system to continue to play out as before, while sustained interventions also act as 'blocks', and stop feedback loops. This is valuable in that it can break a cyclic structure learning problem into several acyclic ones. For instance, one can first check whether X_i affects X_j by performing a sustained intervention on X_i before checking whether X_j also affects X_i. However, this approach will short circuit the natural dynamics that might themselves be a useful source of evidence. As an analogy, one might contrast the punctate (and probably very destructive) intervention of throwing a spanner into a washing machine while it is spinning, to a more systematic sustained intervention in which one holds and turns one of the gears in the mechanism and observes what else moves.

6.5.2 The Role of Theory

The representations we considered in this chapter are abstract in the sense that they do not encode anything specific about the mechanisms involved, notably saying nothing about how the parts of the systems are arranged in space. However, we clearly make use of our knowledge of physical mechanisms when reasoning causally (Ahn et al., 1995; Bramley et al., 2018a). Indeed, much of the recent work in causal cognition has emphasized the 'top down' role of domain theories on causal inferences (Griffiths, 2005; Griffiths and Tenenbaum, 2009; Lake et al., 2015). Such domain knowledge is modelled as a hierarchical prior, affecting people's expectations about what kinds of structure and parameters are plausible in different domains and contexts. For example, basic medical knowledge will tell you that diseases normally cause symptoms rather than the reverse and that medicines can take on the order of hours to work. Basic knowledge of electronics will tell you to expect causal influences to propagate at the sub-second level rather than across the aeons. Thus, while the current frameworks show how causal beliefs can arise without much specific domain knowledge they are also readily compatible with other factors and knowledge playing into the choice of sensible priors. Another way of thinking of specific domain knowledge is as rich functional forms governing the interactions between variables. For example, Bramley et al. (2018a) explore interventional learning about the masses and forces relating objects in simulated physical microworlds. Here learners' sophisticated understanding of how objects with different properties normally interact clearly plays into both the judgements they make and the actions they take in service of learning. Interventional behaviours in this domain start to take on recognizable physical form such as shaking and throwing of objects and one another, but still subscribe to the principles exposed here of creating strong interventional signals that propagate through causal relationships. Learners appear to be able to compare the resulting trajectories in time against expectations simulated from their intuitive representation of physics and so back out the relevant properties. In these domains, our interventions and our interpretations of them become deeply theory laden, depending profoundly on our beliefs about how the domain works in general (Bramley et al., 2017b).

6.5.3 When to Aggregate

This is not to say that there is no place for the CBN in causal cognition. The richer representations explored here are useful for navigating causality at the real time scale. But many other phenomena of interest are too long range and noisy to be identified through close focus on the minutia of change over time, only becoming evident as data is aggregated to a larger scale. Time-sensitive representations allow us to make the leap to this higher level of aggregation. Once we have a robust knowledge of when and where to measure effects we can start to gather evidence on a larger scale where time is increasingly abstracted away and aggregation over instances is possible. For example, reasoning at the level of continuous OU variables might be important for someone to play tennis. But if they then want to decide whether to follow a particular strategy it will be more useful to keep track, over multiple games, of which strategy they pursued and whether it resulted in a win. Essentially, we can use our more detailed and time-sensitive theories to determine the relevant time windows and sufficient measurements for studying the subtler and longer-scale causal relationships we are interested in establishing. Indeed, it is only through this large-scale organized process, supported by our expertise in the fine-grained mechanisms, that humanity has been able to discover the weak relationships between behaviours and medical problems decades later, such as the link between smoking and cancer (Gandini et al., 2008), or the lack of one between vaccinations and autism (Verschuur, 1996). In general then, the best interventional strategy for a learner depends on what they are interested in learning, how much they already know about the domain, as well as the granularity of the available measurements.

6.6 Conclusions

In sum, recent work has begun to shed light on the temporal sophistication of everyday causal cognition. We are aware, as was Heraclitus, that we 'cannot step into the same river twice' (Barnes, 2013). Thus, we must learn to integrate our model-based expectations about how things will play out, with careful interrogatory actions, to step many times into different but related rivers, and so learn and exploit the dynamic causal world in a single lifetime.

7

Time–Event Relationships as Representations for Constructing Cell Mechanisms

Yin Chung Au

7.1 Introduction: Time–Event Relationships

This chapter shows a functional way of perceiving time and events in experimental cell biology and argues that the two kinds of perception are paired to become representational tools for reasoning about cell phenomena and the construction of mechanistic explanations. In cell biology, offering mechanistic explanations remains the mainstream approach, while cell biological complexity means that the practice of constructing explanations for cell phenomena is very complicated. This chapter presents the results of a practice-focused study and argues for the crucial role of pairing the perception of time and the perception of cell events in practice, questioning the assumption that perception of time is merely a background reference. The assumption, while not explicitly stated in the existing literature, is evident in the fact that the epistemic functions of perceiving time–event relationships have not been examined. This chapter will show that experimental cell biologists infer causality from time-related observations by treating time as more than a fixed reference. In some cases, time may even be perceived as a player in cell dynamics, in which the status of causes changes over time. This chapter suggests that in cell mechanism research, the perception of time helps the researcher not only to acknowledge status change but also, importantly, to reveal the patterns of events and thus identify the objects of intervention for future experimental design. The known, unknown, and newly discovered mechanism components are like bricks in models. The use of time–event relationships as representations plays a key role in deciding how to arrange such bricks.

Two initial points should be noted. First, arguing for the importance of perceiving time and events in pairs does not underestimate the contribution of merely perceiving the time points of events, particularly the 'what happens first' observation, to the construction of explanations. In cell mechanism research, acknowledgement and representation of cell events can always be time related. This is because, if cell events can be viewed as either difference-making or production,[1] time is the necessary reference for either measuring the difference or tracking the continuous changes that happen over the course of the processes. While the 'what happens first' observation helps determine difference-making, the perception of continuous temporal patterns of events is used to explore the finer details, such as the trend and the scale of the processes. For example, the recording of 'the chemical concentration first reached the peak and then fell to baseline' shows the timings of two events and is needed for recognising the rise and fall of the concentration. On the other hand, the recording of 'the chemical concentration reached the peak at the 5th minute and then decayed to half every 10 minutes' gives the trend of concentration change. Detailed recording of concentration curves can be used for determining the coefficient of change. These data are normally called 'time courses' by cell biologists and are needed for recognising *in what way* the concentration rises and falls. Cell biologists need both types of data to construct explanations.

Second, since mechanistic reasoning requires synthesis of different scales of time, space and entities,[2] it is very likely also to require representational resources for the convenience of surrogative reasoning. I borrow the term "surrogative reasoning" from Swoyer (1991) and adopt the development by Contessa (2007) of the notion. Based on both accounts, surrogative reasoning in this chapter refers to someone using representations (visual objects, propositions) to learn about the target, i.e. the mechanism of the phenomenon. As the case studies will show, once the researchers have confirmed the evidential status of a piece of data, they use the knowledge that the data is the representation of a specific entity/activity to determine the way of positioning the entity/activity in the existing model. As multi-scale integration is a live

---

[1] The existing literature on causality has discussed extensively the distinction between difference-making (or dependence) and production. However, my position is that the two accounts may be reconciled in some sciences. Illari (2011) argues that in evidence-based medicine, difference-making evidence and evidence of mechanism serve different functions in establishing causal claims. This chapter extends the discussion to the practice of cell biology. Section 7.2 will introduce my emphasis on the importance of determining both kinds of causation in cell biology. Section 7.4.1 will support this emphasis with a case study.

[2] There is a rich body of literature discussing the multi-scale integrative features of mechanisms, the emergence of new properties from integration (Bechtel's 'orchestration' of components also implies an integration), and the scientific benefit of integrating models on multiple scales. For example, see Craver and Darden (2013) and Bechtel (2006) for their views of the multi-scale features, and see Green and Batterman (2017) for the interplay of processes between different scales within the modelling of biological behaviour.

issue in the philosophical literature on mechanism, this chapter will contribute to the literature by explaining how the perception of time serves to help the actual practice of this integration. Namely, by using representational resources constituted by times and events registered at different timescales and in different spatiotemporal frames, cell researchers conduct surrogative reasoning on a multi-scale basis. Such representational resources can be 'moved about' during the process of model construction. At the same time, they symbolise specific patterns of events and the involvement of specific entities. Thus, they can be viewed as operational tools for puzzle solving.

The argument of this chapter has two parts. On the one hand, this chapter argues that perception of time–event relationships is important for determining causality in cell biology, by informing the ways of cross-scale mapping of pathways where events happening on timescales ranging from a few minutes to a few days or more are integrated into a single model, and helping the researcher to identify what properties to observe and intervene in.[3] On the other hand, the chapter argues that time–event relationships are used as representations in mechanistic reasoning about cell phenomena. Section 7.2 will explain how I treat the concepts of causality and pathway in cell biology based on actual practice. This is to provide the ground for arguing for the importance of time–event relationships. Section 7.3 will elaborate on the representational function of time–event relationships in developing mechanistic models. In Section 7.4, three case studies will provide concrete examples to support my arguments elaborated in Sections 7.1 to 7.3. Section 7.5 will summarise the findings and highlight an intriguing observation concerning mechanism diagrams: once the mechanistic models have been constructed, time tends to be removed from the visual representations of the models (mechanism diagrams). Based on this observation, this chapter suggests that the perception of time–event relationships is an intermediate process in the construction of mechanisms. This intermediacy might explain why, in the existing literature, time tends to be viewed merely as a background setting.

7.2 Causality and Pathway in Cell Biology

This section explains how I treat two concepts in the context of cell biological practice. The first concept is causality. This chapter considers that causal relevance and causal productivity coexist in explanations for cell phenomena.

---

[3] Section 7.2 will explain in detail the variety of timescales and the practice of mapping between different scales.

The second concept is the term 'pathway' in cell biology, which refers to a kind of epistemic object that is different from a mechanism. I draw this distinction to emphasise the importance of identifying intersecting points of pathways for constructing mechanisms. Section 7.3 will follow by arguing for the use of time–event relationships in finding these points.

This chapter views causality in cell mechanisms as referring to both what I will call causal 'relevance' and causal 'productivity'. This distinction is derived from the philosophical literature on causality, as first introduced by Hall (2004) although he uses the terms 'dependence' and 'production', and developed by Glennan (2009), who uses the terms 'relevance' and 'productivity' as I do. More precisely, this chapter maintains that cell mechanism research deals with both causal relevance and causal productivity, and that these two causal relations are both involved in mechanism construction. The chapter adopts the general idea of Glennan's (2009) two kinds of causation in biology,[4] while arguing for their coexistence at cellular and molecular levels. In Glennan's definition, causal relevance is "a relationship that can hold between a variety of different kinds of facts and the events that counterfactually depend upon them", whereas causal productivity is "a relation that holds between events connected via continuous causal processes" (Glennan, 2009, 325). A cell biological example is that in the reactions induced by activation of tumour necrosis factor (TNF), the cleavage of cellular proteins is *produced by* enzyme caspase-3, whereas the cleavage of cellular proteins is *relevant to*, but not produced by, the upstream biochemical activation of TNF. Nonetheless, the aim of cell mechanism research is to also reveal the process consisting of connected causalities between TNF activation and protein cleavage, i.e. causal productivity. Thus, cell mechanism researchers need to determine both kinds of causality. Hall's 2004 concept of causal dependence is similar to Glennan's causal relevance, and Hall argues that causal productivity and causal dependence "typically coincide". This chapter maintains that the coexistence of these two kinds goes beyond an accidental coincidence in cell mechanistic explanation. The case study in Section 7.4.1 will explain in detail why both

---

[4] Although Glennan (1996b) also uses the term 'mechanism' in his analysis of causation in physics, I consider his mechanism as very different from the concept of biological mechanism in my study. First, I only deal with mechanisms in biology in this chapter. Second, I follow Craver and Darden (2013), Illari and Williamson (2012), and part of Bechtel (2006) and define a biological mechanism as composed of entities and activities "organized in such a way that they are responsible for the phenomenon" (Illari and Williamson, 2012). The part of Bechtel's view that I do not fully adopt for this chapter is orchestration. Finally and importantly, Glennan's mechanism in physics is something that connects events within a causal relationship, for he uses this to solve the problem of invisible connexion in causal relationships. On the other hand, the term mechanism in my study means expandable mechanistic model that can include causal relations between its component events.

kinds of causality are required and how they are separately determined by
experimental biologists. Here, I summarise why they coexist. To construct
an explanation, cell researchers need to first confirm that an event occurs in
virtue of the occurrence of another event. Then, to construct a *mechanistic*
explanation, cell researchers need to reveal the productive process that connects the two events. Normally, a confirmed biological causal relation tends to
still comprise underlying mechanisms. Given this complexity, cell biologists
approach the completeness of their explanations by seeking to determine
physically continuous causal processes. To ensure both practical convenience
and future fruitfulness, the best research strategy is to recognise the coexistence
of causal relevance and causal productivity.

Further developing ideas of causality, I borrow Hall's concept of causal
locality to emphasise the importance of time–event relationships. Causal locality means that the "causes are connected to their effects via spatiotemporally
continuous sequences of causal intermediates" (Hall, 2004, 225). This chapter
adds that the determination of cell causal locality requires the researchers to
identify the places for probing between known cell events,[5] e.g. deciding where
to posit the hypothetical involvement of further entities and the hypothetical
occurrence of yet-unknown activities. This chapter will show that time–event
relationships play a crucial role in these processes. In Section 7.3, I will
elaborate more on the usefulness of time–event relationships in finding causal
locality.

Here, a related concept should be noted. In the case studies, I will
mention that some events are considered by the researchers to be sufficient
or insufficient causes of some phenomena. This causal sufficiency is defined
within a practical framework, based on how the causality is determined by
experimental inference. This means that when I say that A is considered by
the researcher as a sufficient cause of B, I follow the experimental practice, so
I mean that the occurrence of A in the experiment guarantees the occurrence
of B and makes B observable. This implies that A is, in all the cases studied,
the variable to be tested in a specific experiment. This at the same time implies
that the other variables (potentially) responsible for B are controlled. In the
practice of model construction, the validation of a sufficient cause is enough
for the researcher to add it to the model, where the other variables are subject
to test in further experiments. This is exactly how mechanistic models can
gradually expand, be substantiated, and become complicated.

---

[5] This study neither discusses Hall's detailed definitions of causation nor relies on his locality to
develop the argument. I borrow Hall's conception of locality just to express concretely the
importance of recognising time–event relationships for intervening in such spatiotemporally
continuous sequences.

Now I introduce the second concept of this section: 'pathway' is a terminological nuance that has a unique position in the practice of constructing mechanisms in cell biology. As the case study will show, a pathway[6] is composed of entities, activities and, importantly, causal relations. This chapter considers mechanism components as entities and activities, but I take a step further to argue that 'pathway' is not simply interchangeable with 'mechanism'. The main difference is the inclusion of causal relationships, which makes pathways different from Craver and Darden's mechanisms. Pathways also occupy different places than mechanism components do in the development of mechanistic explanations. Section 7.4.1 will use concrete examples to show that when cell researchers use this term, they clearly refer to something distinct from both mechanisms and mechanism components. With a focus on the practice, Ross (2018) has similarly argued for distinguishing pathways from mechanisms. However, while I generally agree with Ross's view that pathways function differently from mechanisms in explanation, I suggest three points showing that pathway and mechanism are different but not mutually exclusive:

1 Pathways serve as epistemic prototypes subject to interlinking.
2 A mechanism (as used by cell biologists) comprises interlinked pathways.
3 Compared to a mechanism, entities and activities within a pathway are held together cohesively by core causal relationships. This echoes Ross's view that pathway concepts "emphasize the 'connection' aspect of causal relationships" (Ross, 2018, 6).

These three points will be demonstrated by the concrete examples raised in Case 1. The case will also show that when determining a pathway, researchers use both extrapolation and interpolation, based on experience and established knowledge, to decide what new components to posit and where they should be inserted. The extrapolation and interpolation processes are normally explicit in biological papers. The aim of such processes is to find as many direct causal connections as possible between the entities and activities *in a pathway*. The common term 'signalling pathway' used by cell biologists is demonstrative of the important role of causality in pathways. This term refers to a cascade of reactions induced by one or more events, and clearly assumes the existence of an initiating cause. For example, the 'Fas receptor signalling pathway' refers to a group of biochemical reactions that point to cell death or inflammation.

---

[6] It should be noted that pathways normally have vague, flexible boundaries and that the boundaries are not natural but hypothetical. Since pathways figure as component constructs in mechanisms, defining the boundary of pathways is very similar to defining the boundary of mechanisms. The purpose is to idealise the research object. This study takes this position in the vein of Bechtel (2015).

These reactions all start from the activation of Fas receptor (the binding of Fas and its ligand). That is, the event 'Fas activation' is a necessary condition responsible for all the reactions in the pathway, and the research task is to determine what proximate causes to insert between 'Fas activation' and the downstream phenomenon of interest.

As Case 1 will show, cell biologists use pathways as epistemic prototypes of mechanisms and organise them in ways informed by experimentation. The organisation of pathways tends to require cross-scale mapping as described below. Cell mechanisms can always involve both known and unknown entities and activities, which are assumed to have known and unknown effects on each other. The timings of their interactions can be very different, and the scales of the timings are meant to be intertwined. For instance, the time required for seeded cells to grow to adequate numbers for an experiment can range from overnight to four days. Expression of a protein caused by a transfected gene may take up to two days. Activation of an enzyme only takes a few hours, while some other reactions take no more than one hour. Different timeframes are integrated within a single model for explaining a specific phenomenon. Following the existing literature on the integrative nature of biological complexity,[7] this chapter emphasises that mechanisms often synthesise diverse temporal scales. Thus, the composition and organisation of epistemic components, i.e. pathways, require cross-scale mapping in both temporal and spatial aspects. It should be added that this chapter accepts Bechtel's (2015) account for timescale in biological mechanism research. In this chapter 'timescale' means an idealised framework for hypothesising about, and intervening in, the temporal features of cell events in experimentation. Meanwhile, this chapter agrees with the view that biological phenomena can be considered scale free since they do not have well-delineated boundaries in nature. This is the view of Marom (2010), which Bechtel seeks to reconcile with mechanistic explanations. In this regard, pathways do not have natural boundaries, either. They are hypothetical constructs in idealised scenarios for operation.

How does the researcher identify the intersecting points of pathways? The importance of appreciating time courses in such identification is illustrated in both Case 1 and Case 2. In short, observation of how experimental intervention changes the time course of cell events—*not* merely the end

---

[7] With regard to the necessity of integrative pluralism for understanding biological complexity, this study fully accepts the account of Mitchell (2003), despite the terminological difference between 'models' and 'mechanisms' for describing the explanations for biological phenomena. In fact, my previous empirical study (Au, 2016) on a statistically representative population of biological papers has shown that these two terms are used interchangeably by the practitioners. Thus, I view Mitchell's integrative pluralism as very useful in reflecting the reality of cell biological practice.

status—is important for informing the points at which specific pathways converge or diverge. The researcher reasons about the organisation of pathways based on their interpretation of experimental results of the intervention. In other words, the experimental results of intervention must gain *meanings* to represent whether a specific component (entity or activity) of a pathway is the point for other pathways to join. The next section will argue that time–event relationships serve to represent some of the meanings of experimental results.

7.3 Time–Event Relationships are Representations

This section explains why philosophers should pay attention to the representational and practical functions of perceiving time and events in pairs. Briefly speaking, time courses as representations simultaneously help the researcher reveal causal relations between the trends of various behaviours of mechanism components, and serve as objects of conceptual intervention before experimental, physical intervention takes place. The existing literature, including Bechtel's and Marom's views (Bechtel et al., 2014; Bechtel, 2015; Marom, 2010) concerning the timescale of biological mechanisms, which naturally work on the timescale of biological mechanisms and deal with the problems regarding the existence of timescale of such mechanisms, has not dealt with how exactly the perception of time is *used* in the making of mechanistic explanations. The existing literature seems to treat time as something fixed and part of the background setting. I argue that we should refine our philosophical conception of the relationship between the perception of time and other variables so that we can appreciate the meanings and usefulness of time in biological reasoning processes. This use of time is similar to using relationships between chemical concentrations and cell events, where the concentrations are not perceived as fixed references but as players that have relationships with the events. Section 7.1 clarified that cell events can always be related to time. Now I add that the reverse is also true: in the interpretation of experimental results, meaningful recognition of time is often associated with a recognition of events.

Based on the case studies, this chapter will show that functions of signifying meanings of experimental data are given to time–event relationships, as the researcher needs representational resources to develop explanations. The chapter does not deal with the structural similarity between the mechanistic explanation and the phenomenon of interest. The focus is on the practical use of time–event relationships as representations of the behaviour of mechanism components in the process of understanding the causal relations between the behaviours. The examples in Section 7.4 will present how various meanings of experimental data were given to the time–event relationships and that the cell researchers used the time–event relationships as representations that signify the following in the cases studied: the effectiveness of the activation of a nuclear factor, the level of cell mobility,[8] and even the unstable nature of a cause (i.e. the threshold level of a protein for inducing cell events).

By showing the flexible use of time in determining causality, this chapter fills the gap in the existing literature caused by the treatment of time as merely a background setting. While the 'what happens first' observation, for which time is a fixed reference, is essential in the interpretation of experimental results, its contribution can be partial. As stated above, and as Case 1 will show, the understanding of biological complexity requires cross-mapping of pathways and thus identification of the intersecting points of such pathways. Continuous time–event relationships are used for determining the validity of the intersecting points at which some reactions are switched on, some are switched off, and some interfere with some others. Moreover, not all the switch-ons and switch-offs are effective for a long enough time to induce the downstream reactions. At certain points within the time frames of their effectiveness, they may be blocked or even reversed by activities that have different timescales. Hence, working on mechanisms on a cross-scale basis at least means finding out at what points the pathways intersect within specific timeframes.

Moreover, time–event relationships are in fact the objects of conceptual intervention for determining Hall's 'locality' of causation. In practice, time-frames of experimental intervention and observation are normally decided based on existing knowledge of comparable conditions and/or properties of similar entities of other organism systems. Here, time–event relationships are particularly important because they are used to infer where and when to conduct the next intervention. That is, before physical intervention takes place, experimentalists design the intervention process by using time courses as representations of the real pathway components. Here, it should be noted that experimental manipulation of time–event relationships is more feasible than manipulation of space–event relationships, since intervening in a time course is practically more possible than changing the location of a cellular event. In cell biology, changing a time point in a time course can change the

---

[8] When referring to biological level defined by size, this study uses the term 'scale' and deliberately avoids using 'level'. This is to avoid confusion with the level of explanatory elements, which may invite debates on inter-level causation (see Romero (2015)). When the term 'level' is used in this chapter, it refers only to the quantitative level of biochemical substances determined by experiments (such as concentration and amount).

result drastically. This is because many cell mechanisms are regulatory ones that exhibit dependence upon a myriad of time-related factors. This sensitivity of biological mechanisms is a result of Hall's spatiotemporal continuity and makes experimental manipulation of time courses a useful approach to revealing the intersecting points of pathways.

Recognition of and emphasis on time in biological mechanisms are not limited to research that is obviously about time (such as circadian rhythms). By stressing this point, this chapter extends the philosophical discussion on time in biological mechanisms.[9] As suggested earlier in this chapter, the existing view that time serves as a fixed reference does not fully capture the complexity of practice and requires a complementary view. This chapter provides a complementary view by drawing on cases in research fields other than time-dependent biology. The following three sections will discuss three cases in detail. The cases were sampled randomly from a previously established database of journal papers in the apoptosis (programmed cell death) field.[10] This chapter analyses the whole journal paper, including text, data images, and diagrams, to find out how the registration and perception of time–event relationships function in the development of biological arguments. It should be emphasised that the 'mechanisms of interest' in cell biological papers tend to contain groups of reactions. Normally, cell biological papers do not present one single linear causal 'mechanism', but rather plural and dynamic mechanisms that can be broken down into many *pathways*. Thus, this chapter selects only a few of the pathways in each case as the key 'mechanisms of interest' for the convenience of analysis and elaboration. How the pathways reflect the big pictures (mechanisms) also confirms my argument that pathways are epistemic prototypes of mechanisms.

7.4 Cell Biology Cases

The first case is relatively straightforward and demonstrates that both causal relevance and causal productivity need to be determined for approximating

---

[9] Bechtel et al. (2014) have elaborated well on how time is represented by the model of suprachiasmatic nucleus oscillation in circadian rhythm research. However, Bechtel's work does not touch on the role of time-perception in biological research fields other than time-related phenomena such as the biological clock. Furthermore, the study of time being represented by artificially developed mechanistic models seems to imply that the status of time is a fixed reference for perceiving changes. This limitation is exactly the point from which this study seeks to extend the philosophical discussion.

[10] The rules for selecting and sampling journals and papers are detailed in my Ph.D. thesis, Au (2016).

the completeness of explanation. The second case demonstrates the use of a time–event relationship as a representation in validating an extrapolated cause. This is an interesting example in which the time courses of cell events were perceived through observing and registering spatial changes of the cell. The time courses were not directly recorded but were 'felt' in the spatial data. The third case is complicated, where the proximate cause of the phenomenon of interest is the effect of time-dose dynamics. It demonstrates the necessity of multiple layers of causal inference for capturing a part of the complex dynamics. Together, the three cases suggest that the use of perception of time in cell mechanism research is flexible and moves beyond the role of a fixed referential axis.

7.4.1 Case 1: A Time–Event Relationship Informs a Diverging Point

The first case centres on the regulation of nuclear factor kappa B (NFκB) in the scenario that a cell death receptor Fas is activated and the Fas-associated death domain protein (FADD) is recruited.[11] In this case, the phenomenon to be explained is the dual possibilities of Fas activation: it appears to lead to either cell death (apoptosis) or cell survival (inflammation).[12] As the cell cannot be simultaneously apoptotic and surviving, there must exist a switching point where Fas-inducing progression of death is inhibited and converted to inflammation. The key entities are Fas receptor, FADD, NFκB and a few anti-apoptotic ('protective') proteins, as described below.

Although the causal relationship between Fas activation and apoptosis has been established, we cannot simply say 'Fas produces apoptosis', because Fas activation actually produces either apoptosis or inflammation (survival). Thus, Fas activation is merely causally relevant to apoptosis and inflammation, and more proximate causes of either destiny of the cell remain unknown. The hypothesis is that a proximate cause explains the conversion from death to survival. According to established knowledge, inflammation is one of the "apoptosis-independent functions of Fas" (Kreuz et al., 2004, 370), but the molecular mechanisms were unclear. In sum, researchers in this case sought to reveal the cause that produces inflammation under the circumstance of Fas

---

[11] Kreuz et al. (2004). The title of the original paper is "NFκB activation by Fas is mediated through FADD, caspase-8, and RIP and is inhibited by FLIP."

[12] As explained above, for each case I focus on one or a few pathways for the convenience of analysis and elaboration. The necessity of this selection shows the complexity of cell biological research, which usually reveals a group of pathways and arranges them in the developing model.

activation. The researchers sought to discover what entity and activity to insert between Fas, apoptosis and inflammation in the model.

The practitioners' use of 'pathway' in this case demonstrates the unique position of pathway in the practice of constructing explanations. One example is that the researchers mention established knowledge of specific pathways to introduce why they needed to conduct an experiment: "Active caspases cleave components of the NFκB signalling cascade and efficiently inhibit activation of this pathway [...] Therefore, we decided to analyze FasL-induced NFκB signaling and gene induction in cells protected from the apoptotic action of FasL" (Kreuz et al., 2004, 368). The other example is that the researchers manipulated variables within established pathways to discover causes that switch some of the reactions (Kreuz et al., 2004) within the pathways.[13] Above all, the researchers refer to the mechanism diagram as a 'model of Fas signalling pathways', which contains causal relations that connect entities and activities.[14] These examples show that pathways are used as prototypes of mechanisms, but not merely mechanism components without causal relations, at least by the researchers in this case.

The researchers determined the causal productivity (as the key finding) by the experimental observation that the time courses of the target events fit within the expected timeframe. The researchers decided a hypothetical timeframe and designed the timings of observation based on established knowledge. They found that NFκB was activated after the binding of Fas and its ligand TNF. In the scenario of cell death, NFκB activation was transient. In the survival scenario, when Bcl2 (an important survival protein) was overexpressed, NFκB remained active for a longer period, which was enough to induce the downstream cascade of inflammation.

Since Fas activation is causally relevant to *both* survival and death, the determination of causal productivity between Fas activation and the two cell destinies relies on the effectiveness of NFκB activation. Scientifically, success and failure of NFκB activation are represented by the quantitative levels of IκB, an inhibitor of NFκB (Karin, 1999). Loss of IκB level represents NFκB activation, but only a sustained loss of IκB means that the level of NFκB activation is enough to induce inflammation. Thus, the duality of the temporal pattern of IκB level, i.e. being either transient or prolonged, represents the 'branching' of cell destiny when the death receptor Fas is bound with its ligand TNF.

---

[13] See figures 3 and 4 of the original paper (Kreuz et al., 2004), where the experimental data were directly related to the established knowledge of pathways.

[14] See figure 8 of the original paper as cited above.

In this case, the perception of time is clearly associated with the perception of events, i.e. IκB level. In detail, there are two representational time–event pairs: one represents a transient and rapid loss of IκB, which occurs in unprotected cells without Bcl2 expression, and the other represents delayed and prolonged loss of IκB, which occurs in protected cells with overexpressed Bcl2. The researchers concluded that Bcl2 overexpression can convert a pro-apoptotic signalling pathway to one that is anti-apoptotic and inflammatory. Thereby, the researchers found the intersecting point of the two pathways. In the model that explains both cell death and survival, two pathways need to be inserted between Fas activation and the final destinies: NFκB activation, and its upstream pathways. In the model that only explains cell survival, one pathway needs to be inserted between NFκB and survival/inflammation: the group of downstream reactions of NFκB activation. Thus, the organisation of the two different yet related pathways has been greatly informed by the use of time–IκB-level relationships as representations in the surrogative reasoning process.

The time–event pairs are representations of the effectiveness of NFκB activation for inducing inflammation and cell survival. They were used by the researchers to inform the intersecting point of two different pathways: the death pathway and the survival pathway. The intersecting point is also a divergent point of cell fate, as the cell could go in either direction. This is also to say that the relationships between the timings and IκB level changes represent whether NFκB activation in this cellular context counts as a sufficient cause of the phenomena of interest, where the prior condition is that NFκB activation is already known as a necessary cause. As explained above, when I say that NFκB activation is a sufficient cause, I mean that the researchers believe the occurrence of NFκB activation in the context of their experimentation to guarantee the occurrence and visibility of the effects.

This case has shown that time is both a fixed referential axis for measuring status change (increase and decrease of IκB level) and a major element of the representational resources for causal inference. The representational use of time–event relationships provided the experimental data with meanings that are useful in manipulating mechanism components at the conceptual level without physical intervention. This is what I call a surrogative reasoning process, in which the objects of manipulation are various representations of the meanings of data. This case has also shown that the use of continuous relationships between time and IκB level changes informed the divergent point of the pathways. Finally, this case has suggested that cell researchers organise causal pathways to construct mechanisms, suggesting the unique position of pathways in the practice.

7.4.2 Case 2: A Time–Event Relationship Validates an Extrapolation

The second case examines the regulation of cell mobility under different circumstances of cell vitality.[15] My main point here is that in this case, the time–event relationships perceived through visual data are explicitly used by the researchers as representational resources for validating the extrapolated variables. Moreover, and interestingly, while time is the background reference for measuring cell movement, some of the data of the time–movement relationship do not even contain records of timings. This seems to contradict the intuition that time must be recorded to be appreciated. Finally, this case directly supports my argument because the researchers explicitly describe how they deployed the time–event relationships as representational tools in their paper.

This case explores the causes of loss and enhancement of cell mobility in cell death and survival. The key discovery is that an anti-apoptotic protein, human Golgi anti-apoptotic protein (hGAAP), causes an enhancement of cell mobility, where cell mobility signifies cell vitality in the physiological sense. At the beginning of the research process, four established propositions directly relevant to the mechanism of interest were at hand:

- (a) Vital cells exhibit higher mobility than apoptotic cells.
- (b) Calcium ion release from intracellular stores regulates cell mobility via facilitating the turnover of focal adhesion.
- (c) hGAAP enhances calcium ion release from intracellular stores.
- (d) hGAAP inhibits cell death.

The task was to prove the hypothetical mechanism that:

- hGAAP enhances cell mobility and, based on (a) and (d), must enhance cell survival via promoting calcium ions release from intracellular stores, where (c) causes (b).

In other words, the task was to find out whether the extrapolation that adding hGAAP to the pathway of 'calcium ion release causing greater cell mobility' is valid in this cellular context.[16] Given that hGAAP is proven to be causally

---

[15] Saraiva et al. (2013). The title of the original paper is "hGAAP promotes cell adhesion and
[16] migration via the stimulation of store-operated Ca2+ entry and calpain 2".
This chapter repeatedly uses the phrase 'in this cellular context' to emphasise the fact that cell biological experimentation assumes different types of cells to exhibit somewhat different types of behaviour under the same treatment. The normal solution is to conduct a preliminary experiment to confirm that the cell type of interest exhibits the same behaviour as a referential cell type under the same circumstance.

relevant to cell survival, the experimentation was to determine the causal productivity between hGAAP and higher cell mobility. The experimental results of patterns of cell spreading and attachment are used as symbolic representations of cell mobility in this case. The more quickly the cell spreads and repeats attachment-detachment behaviour, the more vital the cell. This pattern can be rephrased as 'higher turnover rates represent greater vitality of the cell'. In both expressions, time is obviously an important (but not necessarily the only) reference for measuring status change.

To prove that it is the way in which time is associated with events that matters in the reasoning process, I raise three examples from the original paper. The first two examples directly demonstrate the representational use of time–event relationships, and the third supports this idea by showing the representational function of other kinds of pairs, such as the pair composed of cell number and concentration of chemical treatment. This is to say that time is perceived as something similar to other parameters, such as cell number and chemical concentration.

The first example is concerned with the perception of the turnover rates of cell adhesion. The rates are perceived through a group of data measuring the durations of adhesion and calculating the constants of adhesion assembly and disassembly rates. These rates, namely the proportional relationships between time and the changes in adhesion assembly and disassembly, are treated as representations of the levels of cell mobility and used for understanding the causal relation between hGAAP overexpression and cell mobility. Second, as normal biological experimentation requires more than one type of result to validate a pathway, the researchers needed to determine the level of cell mobility via another means. They did so by registering the traces of cell migration imaged every five minutes within a given time period (18 hours) (Figure 7.1). The meaning of this trace data is "the speed of cell migration" (Saraiva et al., 2013, 704). Interestingly, it is not the time–distance relation but the number and overall length of the cell traces that visualised the speed. Time is perceived through perceiving the patterns of spatial changes. This meaning enables the time-dependent patterns of cell migration to signify the level of cell mobility. Since 'speed' is the meaning, the perception of time and perception of events must have been used in pairs to signify the meaning.

The researchers explicitly describe the signifying function of the time–event pairs in their discussion (Saraiva et al., 2013, p707). According to their expression, the reasoning process was as follows: three kinds of visualisations of time–event relationships were combined to represent the causal relationship that 'hGAAP enhances cell mobility', i.e. the causal dependence between hGAAP overexpression and high mobility. Another set of visual data of

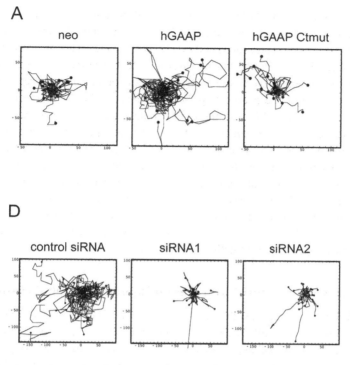

Figure 7.1 Traces of single-cell movement within a spatial frame are used by the researchers to perceive the relationship between time and cell migration. From Saraiva et al. (2013), figure 4A and figure 4D.

time–event relationships represents that the route of this causal pathway is via (c) and (b). This is the hypothesis that awaited proof, as mentioned above. In sum, hGAAP overexpression causally produces the promotion of calcium release, and calcium release in turn causally produces higher rates of focal adhesion turnover. The promoted rates of adhesion turnover eventually produce higher mobility of the surviving cells. Thus, the hypothetical mechanism is now proven by using the various time–event relationships, which came in the form of four different sets of visual data, as representational resources. The extrapolation of hGAAP is now validated.

The third example shows that such a representational function is seen in another type of 'pair': the relationship between the number of cell adhesions and the concentration of EDTA (Ethylenediaminetetraacetic acid) solution. EDTA solution is used in biological experiments to detach adherent cells. Cells susceptible to apoptosis are more likely to be detached by EDTA. Hence, the relationship between EDTA concentration and the number of cells remaining adherent represents cell resistance to EDTA as an external harmful stimulus. The experimental results suggested that hGAAP overexpression promotes this resistance, whereas muting hGAAP has no effect. The researchers know that hGAAP overexpression is the sufficient cause of cell resistance to EDTA treatment, because the relationships between the hypothetical causes (presence and absence of hGAAP) and the effects (attachment and detachment of the cells) served as the representations of the cells' susceptibility and resistance to EDTA respectively. This reasoning could only be completed by combining the EDTA concentrations and the number of adherent cells. Similarly, the afore-mentioned reasoning that 'hGAAP overexpression causes higher cell mobility via promoting calcium release' could only be completed by combining the time points and the number of cell movement traces. Once again, the perception of time, just like the perception of cell adhesion, is most *useful* in the status of being meaningfully paired with other variables.

7.4.3 Case 3: Multi-Layer Causal Inference of Cell Dynamics

The third case exemplifies the necessity of multiple layers of causal inference in the research on cell dynamics.[17] This chapter first uses this case, once again, to demonstrate that perception of time is most useful when coupled with events, and second, to shed light on a direction for future study of mechanistic causality. This future direction should deal with the problem of how time is perceived in complex dynamics. In this case, the role of perception of time is more complicated than in the two examples above. This chapter assumes that such a complicated role is not a unique case in biological research. In a dynamical system, the functions of some independent variables can change over time and thus change their effects on the dependent variables. Specifically, some causes do not remain sufficient throughout the time of experiment. In this case, can one claim that time plays a causal role and is more stable than the time-dependent causes?

For clarity, in this section, I name different causal relations by layers based on their stability during the dynamic processes. A layer-one cause is more stable under different circumstances than a layer-two cause. For example, when event C is caused by event B, and event B *depends* on event A, which changes over time, the relationship between event A and B is named 'layer-one causality', where the relationship between event B and C is named

---

[17] Paek et al. (2016). The title of the original paper is "Cell-to-Cell Variation in p53 Dynamics Leads to Fractional Killing".

'layer-two causality'. Layer-one is more stable than layer-two, because layer-two causality here depends on changes in A over time. It should be noted that I do not deal with whether there is a transitive relation between the two layers, such that if A causes B and B causes C, it follows that A causes C.[18] This case study will focus on an intriguing point that is specific to dynamic systems: the unstable conditions that researchers often face for investigating layer-two causality. This condition is due to the unstable status of event B, because it depends on event A on a temporal basis.

In the process of inferring the two layers, the researchers explicitly used time–event relationships as representations and visualised the numerical form of the relationships as part of the explanation. The case is a non-linear process in which the quantitative definition of the cause (A above) – an excess of the level of a substance over the threshold for inducing the effect – is inconsistent at different time points. This means that the causal relation (between B and C above) only 'exists' within particular timeframes. Hence, the effect (C above) cannot be predicted by simply measuring the level of the substance. Perception of the time course in this case serves in the understanding of the layer-one cause of the 'instability' of the layer-two cause, where the effect, C, induced by the layer-two cause is the phenomenon that needs an explanation. To know where in the mechanism to intervene, the researchers had to first understand the reason why sometimes the layer-two cause (B causing C) is not *valid*.

Through using the time course of threshold changes as a representational tool for reasoning, the researchers concluded that they should intervene in a specific threshold of an entity – but not the entity itself – to produce desired effects. Here, I summarise the scientific process. The researchers sought to develop an explanation for how involvement of p53 (an apoptotic protein) leads to fractional killing of cancer cells when cancer cells are treated with cisplatin, a chemotherapeutic drug that kills cancer cells by inducing apoptosis. In most cases, a p53 level that exceeds its 'death threshold' (B above) is enough to induce apoptosis (C above). However, this causal relationship between B and C was observed unstable in the previous study. Therefore, the researchers aimed to identify where to intervene so that the efficacy of cisplatin can be freed from the limitation observed. The experimental result suggested that the 'right' p53 threshold at specific timings is the object of intervention via manipulating the timing of treatment. The researchers claim that their discovery "underlines the importance of measuring the dynamics of key

---

[18] Transitivity of causation is part of David Lewis's counterfactual theories and has well-publicised debates during the past decades. See Hitchcock (2001) for a summary and counterexamples. Also see Paul and Hall (2013) for detailed case studies that challenge transitivity.

players in response to chemotherapy to determine mechanisms of resistance and optimize the timing of combination therapy" (Paek et al., 2016, 631). Their discussion maintains that "the results described in this paper highlight the fundamental role that timing plays in such [combination therapy of cancer treatment] strategies" (Paek et al., 2016, 639). The researchers concluded that the potential object of intervention is the timing of cisplatin treatment, and the intervention is the addition of other drugs that are cisplatin enhancers. Timing is proven "critical" (Paek et al., 2016, 639) for enhancing the effect.

For the convenience of analysis, I break down the key elements of the causal relations as follows:

- p: fraction of apoptosis in cancer cells under cisplatin treatment. This is the phenomenon to be explained.
- e1: apoptosis under cisplatin treatment. This is the known final effect.
- c2: activation of apoptotic protein p53. This is the known cause that has been proven to be sufficient for producing e1.
- e2: the instability of c2's causal status. This is the effect observed in the process of investigating p.

While the primary task was to explain p, the research process turned in another direction when the researchers discovered a curious phenomenon, e2. Although the causal relation of c2 and e1 has been established, c2 appears to be a 'non-cause' of e1 on some occasions. Given that cisplatin has been introduced to clinical use based on the causal productivity between c2 and e1, this phenomenon seems to contradict the established knowledge of therapeutics, unless a novel mechanism is revealed to be responsible for the unstable causal status of c2. Thus, it became necessary to determine the cause of e2 before the researchers approached the explanation of p. So, what is the condition for c2 to be valid?

The researchers traced the fates of single cells and calculated the number of surviving and apoptotic cells. This was to extract the collective patterns of cell fates under cisplatin treatment. Then the researchers statistically determined the death threshold of p53 level. The conclusion is shown in a conceptual and schematic diagram (Figure 7.2): the threshold changes over time in a non-linear pattern. This explains e2. Combined with further experimental results that revealed the mechanism of upregulating a family of anti-apoptotic proteins, the time course of the death threshold can now be used to form the explanation of p.

My terminology of dividing the causal relationships into two layers can help clarify the scenario here: in the layer-one causality, the 'right' p53 level

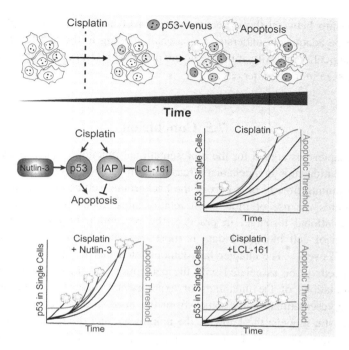

Figure 7.2 The death threshold of p53 level changes over time and thus leads to its seemingly unstable status of being the cause of apoptosis under cisplatin treatment. From Paek et al. (2016), graphical abstract.

that exceeds the death threshold depends on the right timings, whereas in the layer-two causality, the occasional excess of p53 at particular times is responsible for fractional apoptosis (p). The time–threshold relationships can be interpreted as 'sufficient status' and 'insufficient status' for the layer-two cause. Therefore, compared to the above two examples, this case supports the necessity of pairing time and events with more radical evidence: time can be perceived as a participant in deciding the validity of the cause. The use of these relationships as representations of the two types of status clearly facilitated the development of the explanation for p. The researchers concluded that for the efficacy of treatment, cisplatin should be given at specific timings when p53 level can reach the death threshold.

This section has shown that perception of time–event relationships can be used both to describe specific features of mechanism components, as in the previous cases, and to represent a cause. This section has also proposed a division of the causal inference of such a special case into layers, for the relationship between the different causes involved in the final explanation cannot be adequately understood as a chain. Owing to the non-linear nature of biological complexity, multiple layers of causal inference like this example can be seen in many other cases.

7.5 Conclusion

This chapter has argued for the representational function of perceiving time–event relationships in mechanistic reasoning in cell biology. It has suggested that assuming time to be a part of the background setting is limited because it obscures the use of time–event relationships: their function of being a representational resource. In practice, the perception of time in the interpretation of cell biological data is most meaningful when it is associated with cell events. The chapter has demonstrated that the perception of time often needs to be associated with the perception of other variables to form representations of the meanings of experimental data. The importance of such representational pairs comes from the need to arrange the 'bricks' of mechanistic models: the knowns, the unknowns, and the hypotheses. Time–event relationships can inform the intersecting points of different pathways and the places for inserting new components. Meanwhile, for producing desired effects, time–event relationships inform possible objects of intervention. This representational use supports the pragmatic value of mechanism research (in the vein of Craver and Darden (2013)).

This chapter has also argued for the requirement for determining both causal dependence and causal productivity for constructing biological mechanisms. Causal dependence and causal productivity coexist in cell biological pathways and mechanisms. The common laboratory practice is first to determine the causal relevance between two events and then to approximate causal productivity via interpolating and extrapolating variables. This chapter has demonstrated that the representational use of time–event relationships contributes to the validation of interpolated and extrapolated variables.

This chapter closes with an observation concerning mechanism diagrams. This observation could explain why time tends to be considered as merely a background setting and that its relationship with cell events has been neglected. Mechanism diagrams are visualisations of the models constructed, and they serve as both assertive and communicative devices in biological arguments (Au, 2016). In most cases, neither time nor time–event relationships remain visible in the mechanism diagrams published in biological papers. This invisibility suggests that the use of time-event relationships is an intermediate process. Time–event relationships as representational resources tend to fade away from the finalised models, for they function to inform the intersecting points of pathways but do not themselves stand for mechanism components. Therefore, the perception of time–event relationships is arguably an operational and heuristic tool for constructing mechanistic explanations. The tool function of time–event relationships needs to be revealed by a careful analysis of the reasoning process, as it may be invisible in an analysis of the final models. This chapter has completed the analysis and has argued that time–event relationships have important epistemic functions by serving representational roles in the exploration of causality in cell mechanisms.

Acknowledgements

With heartfelt thanks, I would like to put Dr Phyllis Illari at the top of my acknowledgements for her extremely kind help on the completion of the manuscript in the intellectual, technical, and psychological senses. During the final stages, when my personal circumstances did not allow me to spend full energy on the editing, Phyllis offered far beyond what I could ever expected from a colleague and former examiner. Second, my sincere thanks go to Dr Samantha Kleinberg. This chapter was started all because of her organisation of the Time and Causality in the Sciences conference 2017, Hoboken and her invitation to contribute to the edited volume. Third, I have benefited a lot from the conversations during the conference with many scholars, whom I cannot address all in this limited space. Finally, I truly appreciate the support from my colleagues at the Center for Society, Technology, and Medicine at National Cheng Kung University, Taiwan. Specifically, thanks to Dr Bei-Chang Yang and Dr Hsiu-Yun Wang for the intellectually stimulating discussions we had at the T.G.I.F gathering in January 2018. The discussions have made me rethink about the time concept I analyse in this chapter.

8

Causation, Time Asymmetry, and Causal Mechanisms in the Social Sciences

Inge de Bal and Erik Weber

8.1 Introduction

When social scientists want to establish causal claims, they are often confronted with a typical inference problem: given a statistical association between variables X and Y, argue that there is causal relation from X to Y. The origin of the problem is that such a correlation is in principle compatible with three scenarios: a causal relation from X to Y, a causal relation from Y to X, and a common cause Z that influences both X and Y. This inference problem and possible solutions have been discussed by many philosophers and methodologists of the social sciences. For instance, Daniel Steel starts his paper 'Social Mechanisms and Causal Inference' as follows:

> A central problem confronting social research is that an association between two variables can often be explained by the hypothesis that one is a cause of the other or that both are effects of a common cause. (2004, 55)

Steel's article, which we will discuss later on in this chapter, is about how knowledge of social mechanisms can improve causal inference in such situations.

Philosophers and methodologists of the social sciences (and philosophers of the biomedical or behavioural sciences) often discuss a specific subtype of the inference problem sketched above. In this subtype, the challenge has the following structure:

(PC) (1) A correlation between variable X and Y is established in a non-experimental study.

    (2) The goal is to argue that there is a causal relation from X to Y.

    (3) The correlation cannot be explained by a causal relation from Y to X.

(4) To reach the goal in (2) the remaining alternative (a common cause Z which influences both X and Y) has to be excluded.

(PC) stands for problem of confounders. What is important in this subtype is that one of the scenarios that is in principle possible (a causal relation from Y to X) is not considered a viable option by the scientists involved. Their focus lies on excluding common causes.

Our first aim is to clarify the role of considerations regarding time-asymmetry in dismissing the Y to X scenario. This is important to understand the reasoning process that leads to (PC) type situations.

The second aim of this chapter is to draw attention to the fact that in the social sciences the challenge often has a different structure:

(CD) (1) A correlation between variable X and Y is established in a non-experimental study.
(2) The goal is to argue that there is a causal relation from X to Y.
(3) The empirical data support the inverse causal relation (from Y to X) equally well as the causal relation from X to Y that one wants to establish. A common cause Z is also possible.
(4) To reach the goal in (2) a theoretical argument in favour of a causal relation from X to Y has to be constructed.

Situations with this structure are common in the social sciences and constitute a second subtype of the general inference problem. The label (CD) refers to the (problem of) causal direction. What situation social scientists find themselves in ((PC) type or (CD) type) depends on what kind of observational studies have been done.

Our third aim is to shed light on the composition of the theoretical argument mentioned in clause (4) of (CD), and to show that mechanistic evidence plays a crucial role in it.

The structure of this chapter is as follows. In Section 8.2 we provide some background knowledge on methodological options that are available for biomedical and social scientists.[1] In Section 8.3 we clarify the role of time-asymmetry in establishing (PC) type situations (cf. our first aim). In Section 8.4 we present a case (Duverger's laws with respect to political elections) in order to show that situations of type (CD) are bound to be quite common in the social sciences (cf. our second aim). Section 8.5 introduces concepts (viz. 'social

---

[1] The main focus of this chapter is on the social sciences, but the problem of confounders is equally pressing in the biomedical sciences. Drawing parallels with the biomedical sciences is useful in order to reach our first aim, so in Sections 8.2 and 8.3 we will also give biomedical examples.

mechanism' and 'mechanistic evidence') that we need in order to reach our third aim. In Section 8.6 we use Duverger's laws to explain how mechanistic evidence plays a role in addressing (CD) type situations. In Section 8.7 we summarise our findings.

8.2 Aspects of Causal Inference in the Biomedical and Social Sciences

8.2.1 The Impact of Ethical Regulations

Biomedical scientists investigating the causes of diseases are restricted in their methodological choices by ethical regulations. Randomised experiments with the target population (i.e. humans) are the best experimental method for establishing causal relations in the biomedical sciences. As John Dupré wrote:

> A decisive test of whether smoking causes heart disease, then, would be to take a large sample of human infants randomly selected from the human population, divide them into two equal groups, and force one group to smoke for the rest of their – no doubt abbreviated – lives. (1993, 202–203)

Randomised experiments are usually not among the available options for ethical reasons: they may cause physical harm to the experimental subjects, as in Dupré's example. Ethical regulations (e.g. the Nuremberg Code) forbid this. As an alternative for these unethical experiments, biomedical scientists perform randomised experiments with animals (which are of course not uncontested either) or do observational studies. In an observational study there is no intervention. So the experimenter cannot cause physical harm to the participants, because the participants decide themselves about how they behave.[2]

In the social sciences ethical restrictions similar to those in the biomedical sciences apply. For instance, in clause 12.1(f) of the latest version of the Code of Ethics of the American Sociological Association (June 2018) we read:

> In their research, sociologists do not behave in ways that increase risk, or are threatening to the health or life of research participants or others.[3]

As a result of these ethical restrictions, social scientists often have to rely on observational studies.

---

[2] Nevertheless, informed consent is normally required, because of privacy concerns: the researchers collect confidential data about the participants.

[3] See www.asanet.org/sites/default/files/asa_code_of_ethics-june2018.pdf .

8.2.2 Prospective and Retrospective Studies

There are different types of observational studies. For the development of the argument in this chapter we need a characterisation of two types: prospective studies and retrospective studies. Prospective studies are characterised by the fact that, at the start of the study no one in the experimental group nor the control group exhibits the effect under investigation. All occurrences of the effect are in the future relative to the time the sample was selected (Giere, 1997, 225). Suppose we want to investigate whether smoking leads to lung cancer. We follow 10,000 people with healthy lungs (no indications of lung cancer) for 10 years and passively monitor their smoking habits. We do not tell the participants who has to smoke and who not. They decide themselves. So we are doing an observational study, not a randomised experiment. Moreover, the study is a prospective one because we start with people that do not have the effect that is investigated (lung cancer).

In a retrospective design, the experimental group is a sample of people who already have the effect under investigation, for instance breast cancer (see Giere, 1997, 231–235). This group is compared to a control group in which the investigated effect is absent. The aim of the study is to find out whether the suspected causal factor (the use of oral contraceptives in Giere's example) occurs more frequently (in a statistically significant way) in the experimental group than in the control group. In the example it turned out that there was a correlation between 'long-term use of oral contraceptives' (X) and 'breast cancer' (Y).

Observational studies (including prospective and retrospective ones) have an important disadvantage compared to randomised experiments. It is possible that, after analysing our data, we find out that there are more correlations than the one we are primarily interested in. In the first example above, it is possible that, besides a correlation between smoking (X) and lung cancer (Y), we also observe a correlation between drinking coffee (Z) and lung cancer (Y); and on top of that a correlation between smoking (X) and drinking coffee (Z). This set of three correlations can be explained by three scenarios:

(a) Smoking is a positive causal factor for both developing lung cancer and drinking coffee. Drinking coffee is not a positive causal factor for developing lung cancer: their correlation is explained by the presence of a common cause (viz. smoking).

(b) Drinking coffee is a positive causal factor for both developing lung cancer and smoking. Smoking is not a positive causal factor for developing lung cancer: their correlation is explained by the presence of a common cause (viz. drinking coffee).

(c) Both smoking and drinking coffee are a positive causal factor for developing lung cancer.

It is scenario (b) that has to be excluded if you want to argue that smoking causes lung cancer. How this can be done is not an issue in this chapter, though we will explain this partially in Section 8.5. What is important now is that by doing a randomised experiment, we can avoid the tricky situations in which we have three correlations. Because of the random assignment and experimental manipulation, the purported cause variable X will not be correlated to any other variable. Whatever Z is, there will be no correlation between X and Z. This is why correlations observed in a randomised experiment cannot be explained by a common cause.

8.3 Time-Asymmetry and the Problem of Confounders

(PC) type situations constitute an inferential challenge, but they also incorporate a 'relief': a causal relation from Y to X is not a viable explanation of the correlation. In this section we clarify where this relief comes from, and that time-asymmetry is crucial. More precisely, we will argue that prospective and retrospective studies have a built-in time-asymmetry which leads to (PC)-type situations rather than (CD)-type situations.

8.3.1 Time-Asymmetry in Prospective Studies:
A Biomedical Example

We develop the example of Section 8.2.2 further. We follow 10,000 people with healthy lungs (no indications of lung cancer) for 10 years and passively monitor their smoking habits. Suppose that of our 10,000 people, 4,820 regularly smoke, while the others never smoke. After those 10 years, we observe that the amount of lung cancer cases is significantly higher in the first group ('experimental' group) than in the second group (control group). We may conclude that there is a correlation between smoking and lung cancer.[4] The correlation as such cannot establish a causal relation between smoking and lung cancer (see Section 8.2.2). It does suggest a causal connection between properties 'smoking' and 'lung cancer', but leaves two options open. We have a (PC) type situation:

---

[4] For the sake of argument, we assume that our observational study is perfect: there is no selection bias or any other problem.

(1) A correlation between 'smoking' (X) and 'lung cancer' (Y) is established in a non-experimental study.

(2) The goal is to argue that there is a causal relation from 'smoking' (X) to 'lung cancer' (Y).

(3) This correlation cannot be explained by a causal relation from 'lung cancer' (Y) to 'smoking' (X).

(4) To reach the goal in (3) the remaining alternative (a common cause Z which influences both X and Y) has to be excluded.

Why do we have a (PC)-type situation here, and not a (CD)-type? Why are there two options rather than three? These are the crucial questions now.

It is important that we observed the correlation by studying people who smoked and later on developed lung cancer, not the other way round. If we were interested in the other relation, namely what having lung cancer does to smoking habits, we would have set up another observational study. We would have e.g. passively followed people with lung cancer and without lung cancer and investigated on whether there is a difference in whether they start smoking. In our hypothetical study, the smoking occurs before the lung cancer. We chose to follow people who smoke and monitor their development of lung cancer compared to people who do not smoke. In order to explain the results (the correlation) of the study *as it was designed* by means of a causal influence from lung cancer to smoking, we would have to assume that there is a (biological or psychological) mechanism by means of which lung cancer can have a retroactive (backwards-in-time) effect on previous smoking behaviour. That would be very implausible. This means that a causal relation from Y (lung cancer) to X (smoking) is excluded as a valid explanation of the results of the study with this design.[5]

8.3.2 Time-Asymmetry in Prospective Studies: A Social Science Example

As a social science illustration we consider the study on the relation between parental separation and smoking initiation as it is presented in Kirby (2002). One of the claims put forward is:

---

[5] This does not entail that lung cancer cannot cause smoking (e.g. by means of forward-operating psychological mechanisms that make people diagnosed with lung cancer smoke more – for instance because they don't care anymore about health – after the diagnosis). But as already mentioned: in order argue for this, you have to design a different study.

Parental separation (X) is a positive causal factor for smoking initiation at
adolescent age (Y).

Because of the potential harm of parental separation (for parents as well as for
children) this causal relation cannot be investigated by means of a randomised
experiment (in which we would force some couples to divorce). For ethical
reasons, only observational studies can be done in order to gather empirical
evidence to support this claim.

Kirby takes his data from the National Longitudinal Study of Adolescent
Health (see Kirby, 2002, 34 for details). American high school students were
interviewed twice with approximately a one-year interval. Kirby's sample
contains students who at wave 1 (the first interview) reported to live with two
parents and never having tried cigarettes. His 'experimental' group consists of
those students who experience divorce or separation by wave 2 (the second
interview). The control group consists of students who still live with two
parents at wave 2. For both groups, the relative frequency of smokers at wave
2 was calculated. The results showed a clear correlation: in the experimental
group about 27 per cent started smoking, in the control group only about 16
per cent. Because of the prospective nature of the design (the initial sample
contained non-smokers only, and the smoking initiation was monitored) this
correlation cannot be explained by a causal relation in the opposite direction:
the fact that an adolescent smokes at a given time t cannot be a cause of the
fact that his/her parents divorced in the year before t.

8.3.3 Conclusions

The time-asymmetry that is built-in in prospective designs is crucial for obtain-
ing a (PC)-type situation. Without this time-asymmetry the 'no backward
causation' line of argument that we have used cannot be applied. And if this
argument cannot be applied, the situation is different: we have a (CD) type
instead of (PC) type.[6]

Not only prospective designs, but also retrospective designs lead to (PC)
type situations, because they also have built-in time-asymmetry. For instance,
in the example in Section 8.2.2 this correlation cannot be explained by
a causal relation from breast cancer now to use of oral contraceptives in
the past.

---

[6] Or a situation of a third type, but definitely not (PC), because there it is assumed that a causal
relation from Y to X is already excluded as a possible explanation of the empirical results.

8.4 Why (CD)-Type Situations are Common
in the Social Sciences

8.4.1 Introduction

The aim of this section is to argue that (CD) type situations are bound to appear in the social sciences. This is the second aim that we have put forward for this chapter. In the paper mentioned in our introduction, Steel states that the search for causal relations in the social sciences

> ... leads directly to the thorny problem [of] making causal inferences without the aid of experiment. The classic obstacle to such inferences is that a probabilistic dependence between two variables might be explained either by one variable being a cause of the other or by the existence of a common cause of both. We can call this the problem of confounders, where the term confounders refers to common causes, often unmeasured, that might explain an observed correlation. (2004, 59)

As should be clear from Sections 8.2 and 8.3, we agree that this is a classical problem situation in the social sciences, and we have argued that it is connected to the use of prospective and retrospective designs. However, we should not forget that there are types of situations that are equally thorny (or even thornier) and equally classical (i.e. they also occur frequently): (CD)-type situations. That they are thorny will become clear in Section 8.6. Here we give an elaborate example and explain why they are common. Our example uses the causal relations put forward by the French political scientist Maurice Duverger. Section 8.4.2 contains the required background information on the case. In Section 8.4.3 we show that Duverger was in a (CD)-type situation, while Section 8.4.4 explains why such situations are bound to be common in the social sciences.

8.4.2 Duverger on Electoral Systems

The French political scientist Maurice Duverger became famous in the 1950s for his work on the relation between electoral systems and the number of political parties. The key propositions are the following (Benoit, 2006, 70; Duverger, 1959, 217 and 239):[7]

(D1) The simple-majority single-ballot system favours the two-party system.
(D2) The majority system with a second-round runoff favours multi-partism.
(D3) Proportional representation favours multi-partism.

---

[7] The original book in was in French: Duverger (1951).

In simple-majority single-ballot systems there is one member of parliament to be elected in each voting district. The candidate who gets more votes than any other candidate is elected (even if there is no majority, i.e. the candidate's score is less than 50 per cent). Duverger considers two other systems: the majority system with a second-round runoff (if no candidate receives more than 50 per cent of the initial votes, there is a second round with the top-two candidates) and proportional representation (multiple members of parliament for each district; seats allocated based on percentage of votes for each political party). A two-party system is characterised by the fact that two parties have almost all the seats (e.g. 90% or more). In election studies, political parties that have less seats than a certain threshold (e.g. 5 per cent of the total number of seats) are often called 'invisible'. In this terminology, a two-party system usually has only two visible political parties. In a multi-party system there are usually three or more visible political parties.

Before we have a look at Duverger's evidence and arguments, it is good to reflect on the meaning of the claims. Every causal claim is implicitly or explicitly about a domain. For example, if we say that smoking causes lung cancer, we usually mean that smoking causes lung cancer in *human beings* (and not in e.g. elephants or ravens; the latter claims may be true, but less interesting, if interesting at all). In the case of Duverger, the domain that he and his fellow political scientists are talking about is the set of all democratic countries.

Taking this into account, the meaning of causal claim (D1) can, in our view, be explicated as follows:

> If in all democratic countries a simple-majority single-ballot system would be installed, there would be more countries with a two-party system than if in all democratic countries the voting system would be set to proportional representation.

Similarly, (D3) can be explicated as follows:

> If in all democratic countries a proportional representation system would be installed, there would be more countries with a multi-party system than if in all democratic countries the voting system would be set to simple-majority single-ballot.

These explications are interventionist in a generic sense: the idea is that by intervening on the cause variable one can bring about a change in the effect variable. However, we do not commit ourselves to a specific interventionist theory of causation.[8]

---

[8] The best known interventionist account is Woodward (2003, chs 2 and 3). But Steel (2008, ch. 2) also contains an interventionist account of causation.

The causal relation between smoking and lung cancer (like most causal relations in the biomedical sciences) is probabilistic, not deterministic: smoking is *not* a sufficient cause of lung cancer. There are people who smoke and do not develop lung cancer. However, we have good scientific evidence to believe that smoking *favours* lung cancer: if all people in the world would smoke, there would be more lung cancers than if no one in the world would smoke. Analogously, the causal relation between electoral systems and number of political parties is a non-deterministic, probabilistic one. Not every democratic country with a simple-majority single-ballot system has a two-party system (Canada and India are exceptions); nor is it the case that all countries with proportional representation have a multi-party system (Austria was an exception until 1990).

After putting forward the causal claims, Duverger first gives correlational evidence. He states that, apart from some rare exceptions, dualist countries use the simple-majority vote and simple-majority vote countries are dualist (1961, 248). Dualist countries are countries with a two-party system. So he claims the following:

(D4) The simple-majority single-ballot system is positively correlated with a two-party system.

This claim is based on the status in the UK and the British Dominions (such as Canada and Australia) and its dominions and on the status in Turkey and the USA (1961, 148). Based on the status in Belgium, Denmark, France, Ireland, Italy, Norway, Switzerland, Sweden, The Netherlands, West Germany, and some historical cases (e.g. the German Weimar Republic, republican Spain) he claims that there is a second correlation (1961, 276):

(D5) Proportional representation is positively correlated with multi-partism.

8.4.3 Why Duverger Was in a (CD)-Type Situation

The situation in which Duverger found himself after performing his comparative research is a (CD)-type one. It can be schematically represented as follows:

(1) A correlation between variables 'electoral system' (X) and 'number of visible political parties' (Y) is established in a comparative study.
(2) The goal is to argue that there is a causal relation from X to Y.
(3) The empirical data support the inverse causal relation (from Y to X) equally well as the causal relation from X to Y that one wants to establish. A common cause Z is also possible.

(4) To reach the goal in (2) a theoretical argument in favour of a causal relation from X to Y has to be constructed.

The evidence meant in claim (1) and the goal in claim (2) were clarified in Section 8.4.2. Let us look closer at claim (3). Correlation (D5) is evidence for for causal claim (D3), but it supports the following causal claim equally well:

(D6) Multi-partism (as a result of a given election) favours proportional representation (as an electoral rule in future elections).

The important feature is that, because the correlation is atemporal, it supports a causal relation from X to Y but also from Y to X.

Note that (D6) should not be confused with the following:

(D7) Multi-partism (as a result of a given election) favours proportional representation (as an electoral rule in the given election).

Suppose that after an election in a simple-majority single-ballot system two political parties are represented in the parliament of a country. Then you intervene on this: you force each party to split up in three smaller ones. We guess that no one believes that such an intervention could change the electoral laws under which the elections took place: that would be backward causation. This intervention may have an effect on electoral laws at future elections (cf. (D6)), but not on electoral laws valid at past elections.

As to claim (4), it is clear that Duverger was aware that the 'general coincidences' (this is how Duverger labels (D4) and (D5)) are not enough to establish a causal relation. He developed a theoretical argument that will be discussed in Section 8.6.

8.4.4 Simultaneity in Observational Studies

Why are (CD)-type situations bound to be common in the social sciences? Many observational studies investigate a possible correlation without any temporal direction in the designs. Therefore (CD)-type situations should be considered as 'classical' as (PC)-type situations. Let us first give some textbook examples.[9]

A scientist who is interested in mental well-being may investigate a sample of high school students and find out that there is a negative correlation between depression score and the number of social events attended per week. If this turns out to be the case, a (CD)-type situation arises: the correlation indicates

---

[9] This example and the next one are based on exercises in Hinkle et al. (1988, ch. 7).

that maybe depression leads to less social activities; but the correlation equally well supports a causal relation from low level of social activity to depression. The reason why these two possibilities remain open is that there are no time waves in this study: depression score and social activity level are measured simultaneously.

A second example is the relation between marital satisfaction (X) and job satisfaction (Y). Suppose you find a positive correlation. Because X and Y are measured simultaneously, the correlation can be explained by a causal relation from X to Y or from Y to X. A common cause Z is also possible.

Such 'purely correlational' studies, which are clearly distinct from prospective and retrospective studies, are quite common in the social sciences. Let us illustrate this by means of some recent studies on the connections between smartphone usage and depression.[10] As a result of the rise of smartphones this is a very popular topic. Many correlational studies are available, such as Demirci et al. (2015) and Elhai et al. (2016), all reporting results of observational studies measuring the two variables simultaneously. The results are similar: the studies associate higher depression levels with higher levels of smartphone usage. These researchers seem to be aware that their data do not allow them to argue for a specific causal direction between these variables. Based on the data, they cannot determine whether high smartphone usage leads to depression or vice versa. The authors are nevertheless interested in finding causal relations. This is clear from the theoretical reasoning and reflection that they provide about the possible causal directions. They also compare their reasoning to that of others. So while they are careful to respect the limitations of their data, they clearly look for ways to draw some causal conclusions.

8.5 Social Mechanisms and Mechanistic Evidence

8.5.1 Why Mechanisms?

The standard way to handle (PC)-type situations is 'conditioning on potential confounders'. We illustrate it by means of the examples from Sections 8.3.1 and 8.3.2. Suppose that, besides a correlation between smoking and lung cancer, we also observed a correlation between drinking coffee and lung cancer. We furthermore know that there is a correlation between smoking and drinking coffee. This gives us three possible scenarios:

---

[10] We thank an anonymous referee for suggesting this.

(A) Smoking is a positive causal factor for both developing lung cancer and drinking coffee. Drinking coffee is not a positive causal factor for developing lung cancer: their correlation is explained by the presence of a common cause (viz. smoking).

(B) Drinking coffee is a positive causal factor for both developing lung cancer and smoking. Smoking is not a positive causal factor for developing lung cancer: their correlation is explained by the presence of a common cause (viz. drinking coffee).

(C) Both smoking and drinking coffee are a positive causal factor for developing lung cancer.

It is scenario (B) that has to be excluded. This can be done by conditioning on 'drinking coffee' as a variable. This means that we check whether the correlation between smoking and developing lung cancer still holds if we take only non-coffee drinkers or only coffee-drinkers into account. If the correlation between smoking and developing lung cancer remains in at least one group, we know that the correlation is not due to the presence of the common cause of drinking coffee. So scenario (B) is excluded. Drinking coffee is not a common cause (Z) that can explain the correlation between X and Y.

In the study discussed in Section 8.3.2, Kirby checked three potential confounders that could be common causes: household income, parental education, and smoking status of parents (2002, 61). However, there are many other variables that can and should be considered in the same way. Hence, the procedure of conditioning on potential confounders has its limitations. If we would be able to run a check for all potential common causes, it would be a perfect procedure. However, it is practically impossible to test all potential confounders. It is common scientific practice to test for some potential common causes (selected based on theoretical considerations and past experience) and then consider the causal relation to hold until proven otherwise. Steel formulates this as follows:

> [S]ocial scientists are rarely able to measure all potential common causes.
> Indeed, the inability to exhaustively consider all potential common causes is a
> basic element of the problem of confounders, to which mechanisms are being
> considered as a partial solution. (2004, 63)

Steel's view is that, because of this non-exhaustiveness, mechanisms can play an evidential role here.

He distinguishes two possible roles for mechanisms in relation to the problem of confounders. The first possible role is a negative one: if we do not find a plausible mechanism linking two variables, we can conclude that the correlation between them is spurious (i.e. there is a common cause). Steel thinks that this negative role does not work, because we can always find plausible mechanisms. The second possible role is positive: if we find a mechanism for which we have good evidence we can conclude that there is a causal relation between the two correlated variables.[11] Because the second role does work, Steel is convinced that 'mechanisms significantly aid causal inference in the social sciences' (2004, 63).

We think that Steel is right in his assessment of the evidential role of mechanisms in (PC) type situations. In Section 8.6 we argue that mechanistic evidence plays a crucial role in dealing with (CD)-type situations. Before we engage in this, we clarify what social mechanisms are and what mechanistic evidence is.

8.5.2 What is a (Social) Mechanism?

First, what is a mechanism? For the purposes of this chapter, the following characterization by Stuart Glennan (in his book *The New Mechanical Philosophy*) is adequate:

> A mechanism for a phenomenon consists of entities (or parts) whose activities and interactions are organized so as to be responsible for the phenomenon (2017, 17).

As Glennan emphasises, this is a minimalistic characterisation: not much is required for something to be a mechanism. That is interesting for us here, because social mechanisms can then be characterised as a subcategory of mechanisms in general.

Steel characterises social mechanisms as follows:

> Social mechanisms in particular are usually thought of as complexes of interactions among individuals that underlie and account for aggregate social regularities. [...] But there is more to social mechanisms than just individual interactions: typically, the individuals are categorized into relevantly similar groups defined by a salient position their members occupy vis-à-vis other members of the society[.] (2004, 57–58).

Or briefly:

> Social mechanisms are complexes of interacting individuals, usually classified into specific social categories, that generate causal relationships between aggregate-level variables (2004, 59).

---

[11] Steel develops an account of how to provide evidence for causal mechanisms which he calls 'process tracing' See Steel (2004, 2008, chs 5 and 9).

This is a definition of a specific type of mechanisms which fits the general idea of Glennan: the entities are human individuals, the phenomenon that is produced is a causal relationship between aggregate-level social variables.

When we talk about social mechanisms in the remainder of this chapter, we have Steel's characterisation in the background.

8.5.3 What is Mechanistic Evidence?

Now that we have clarified what (social) mechanisms are in this chapter, we can also clarify what mechanistic evidence is. Mechanistic evidence is bottom-up evidence: it consists in information about the behaviour of the parts that is used to support causal claims about the system as a whole. In the social sciences, providing mechanistic evidence consists in using knowledge about the behaviour of individuals to support causal claims about societies as a whole. In the biomedical sciences, it consists in using knowledge about human individuals and/or parts of the human body to make claims about populations of individuals.

8.6 Mechanistic Evidence in (CD)-Type Situations

We are now ready to tackle the third aim of this chapter: to shed light on the composition of the theoretical argument mentioned in clause (4) of (CD), and to show that mechanistic evidence plays a crucial role in it. In Section 8.6.1 we explain how Duverger brings in causal direction by means of mechanistic evidence, while in Section 8.6.2 we draw general lessons from the case.

8.6.1 Duverger's Mechanistic Evidence

Duverger brings in causal direction by invoking what he calls 'the mechanical effect' and 'the psychological factor':

> The mechanical effect of electoral systems describes how the electoral rules constrain the manner in which votes are converted into seats, while the psychological factor deals with the shaping of voter (and party) responses in anticipation of the electoral law's mechanical constraints. (Benoit, 2006, 72)

What Duverger points at are social mechanisms as defined by Steel (see Section 8.5.2).

Let us look at the first mechanism in more detail. When polling stations are closed, votes are counted. This is done by people (with all kinds of technological assistance) that perform certain roles in the electoral system. These roles are the 'social categories' which Steel refers to, e.g. 'chairman of totalisation office' or 'secretary', or 'assessor' in such offices. The interaction between all these people which count and process votes in a predetermined, highly structured way leads to the proclamation of a result (again by an individual with a specific social role) in terms of seats in the parliament.

This whole process is an input-output system in which votes are processed according to certain electoral rules and result in a distribution of seats. There is always a certain mismatch between share of votes and share of seats:

> The mechanical effect of electoral systems operates on parties through the direct application of electoral rules to convert votes into seats. In the mapping of vote shares to seat shares, some parties – almost always the largest ones – will be 'over-represented,' receiving a greater proportion of seats than votes. Because this mapping is a zero-sum process, over-representation of large parties must create 'under-representation' of the smaller parties. (Benoit, 2006, 73)

However, in simple-majority single-ballot electoral systems, application of the electoral rules leads on average to higher over-representation of large parties.

An important aspect of this mechanism is that it is causally directed. The people involved do not count seats and convert them into votes. They count votes and convert them into seats. And electoral rules are also an input of the process, not an output. The rules are in the heads of the individual (and implemented in the computer programmes they use). They are in no way an output of the vote processing system.

About the second mechanism, Benoit writes:

> Duverger's psychological effect comes from the reactions of political actors to the expected consequences of the operation of electoral rules. The psychological effect is driven by the anticipations, both by elites and voters, of the workings of the mechanical factor, anticipations which then shape both groups' consequent behavior (Blais and Carty, 1991, 92). Under electoral rule arrangements that give small or even third-place parties little chance of winning seats, voters will eschew supporting these parties for fear of wasting their votes on sure losers. Political elites and party leaders will also recognize the futility of competing under certain arrangements, and will hence be deterred from entry, or motivated to form coalitions with more viable prospects. (2006, 74)

Again this is a social mechanism in the sense of Steel: individuals with certain roles ('party leader', 'voter') behave and interact in certain ways. The mechanism rests on two assumptions about how relevant behaviour is determined:

(1) What voters do in the polling station is influenced by their knowledge of the electoral system and the degree to which it favours large parties.
(2) What party leaders do in terms of pre-election coalitions is influenced by their knowledge of the electoral system and the degree to which it favours large parties.

Again, there is a clear temporal and causal order: the electoral rules exist, they influence what is in the mind of voters and party leaders. And the opinions of voters and party leaders determine their behaviour.[12]

8.6.2 General Lessons

A first important question is: how can social mechanisms help? The answer is that they are causally directed. This can be clarified by considering 'Coleman's Boat'. This diagram in Figure 8.1 originates in the work of the American sociologist James Coleman, and has an important role in his book *Foundations of Social Theory* (1990). It has been used by many social scientists and philosophers of social science since then. Here is our version of it:

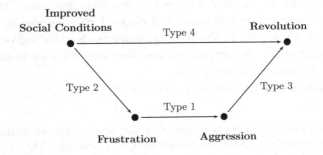

Figure 8.1 Coleman's Boat.

---

[12] Most of the theoretical reflections in the articles about smartphone usage and depression we mentioned in Section 8.4.4 can also be characterised as reasoning about possible causal mechanisms.

This version incorporates one of Coleman's examples (1990, 10). The macro-level causal relations (the deck of the boat) are usually called type 4 connections (see e.g. Little, 2012). They are the ones that have to be established in (CD)-type situations. Type 2 connections are formative: they are about how macro-level institutions influence the beliefs and desires of individuals in a society. Type 1 connections are causal relations within an individual's agency (how beliefs and desires influence actions). Type 3 connections are aggregative: they are about the macro-level phenomena generated by the behaviour of individuals. Coleman's Boat is a graphical representation of the kinds of causal relations that are generally assumed to be operative in social mechanisms. They put constraints on the kind of story you can tell, but it has an advantage: by working with type 1, 2, and 3 connections, you create a causal chain from one macro-variable (X) to the other one (Y).

It is useful to add two 'warnings'. First, it is important that social scientists provide evidence for the social mechanism that they propose (and thus for the type 1, 2, and 3 connections that they claim to exist). Duverger remained rather speculative. Political scientists have done empirical research with respect to Duverger's psychological factor. For instance, it has been shown that the mean vote share of 'third parties' (i.e. all parties except the top two nationwide taken together) is 29 per cent in proportional representation systems, while it is only 17 per cent in simple-majority single-ballot systems (Blais and Carty, 1991). This evidence is crucial in establishing that the mechanism really exists.

The other warning is that there may be a mechanism in the opposite direction. In that case, mechanistic evidence would lead us to accept causal claim (D3) as well as the converse causal claim (D6) – while the backward causation in (D7) is of course still rejected. Some political scientists have argued that there is such a reverse causal relation. For instance, Josep Colomer writes:

> The emphasis on this line of causality does not deny, of course, that existing electoral systems offer different positive and negative incentives for the creation and endurance of political parties. However, precisely because electoral systems may have important consequences on shaping the party system, it can be supposed that they are chosen by already existing political actors in their own interest. (n.d., 1)

So Colomer argues that there is a second causal relation, in the opposite direction. Kenneth Benoit (in different paper than the one we have used till now) agrees with this. In his view, a complete theory of elections must look at...

> ... the dynamic interplay between the forces exerted by political institutions on political parties and the forces exerted by parties to reshape institutions. (2007, 370)

8.7 Conclusions

When discussing the issue of correlation versus causation, Hinkle et al. write:

> The order of the variables' occurrence may indicate whether or not a cause-and-effect relationship is possible. Suppose that the X variable consists of treatments, such as various levels of drugs dosage, and that the Y variable is the performance of a rat running a maze. X cannot affect Y until the drug is administered. If the rat runs the maze before receiving the drug dose, the dose cannot have affected the rat's performance (1988, 139)

The qualifier 'may' in the first sentence of this quote is important: sometimes there is a time order that can be used, sometimes not. Hence, we have distinguished between two types of situations in which social scientists may find themselves when they are in the process of developing an argument for a causal claim. The first type can be schematically represented as follows:

(PC) (1) A correlation between variable X and Y is established in a non-experimental study.

(2) The goal is to argue that there is a causal relation from X to Y.

(3) The correlation cannot be explained by a causal relation from Y to X.

(4) To reach the goal in (2) the remaining alternative (a common cause Z which influences both X and Y) has to be excluded.

This type of situation was illustrated by means of James Kirby's study of the causal relation between parental separation and smoking initiation. These situations are 'classical' because they typically arise from prospective designs and other observational studies with a built-in temporal asymmetry (such as retrospective designs).

The second type can be schematically represented as follows:

(CD) (1) A correlation between variable X and Y is established in a non-experimental study.

(2) The goal is to argue that there is a causal relation from X to Y.

(3) The empirical data support the inverse causal relation (from Y to X) equally well as the causal relation from X to Y that one wants to establish. A common cause Z is also possible.

(4) To reach the goal in (2) a theoretical argument in favour of a causal relation from X to Y has to be constructed.

This type of situation was illustrated by means of Duverger's claim on the effect of electoral systems on the number of visible parties. They also deserve the label 'classical', because this is what occurs when the data in an observational study have no clear ordering in time.

In Section 8.6 we have clarified how mechanistic evidence can play an evidential role when dealing with (CD) type situations. With respect to (PC)-type situations Steel has argued that mechanisms can be a significant aid (see Section 8.5.1). In (CD)-type situations a stronger claim seems warranted: mechanistic evidence is necessary, because there is no other instrument available. If the correlation withstands scrutiny by conditioning on potential confounders, we still have no argument to prefer a causal relation from X to Y above a causal relation from Y to X.

Acknowledgments

We thank Roxan Degeyter and the two referees for their comments on the first version of this chapter. We also thank Phyllis Illari for advertising the main points in this chapter under unusual circumstances at TaCitS2017.

9

Temporalization in Causal Modeling

Jonathan Livengood and Karen R. Zwier

Causal influence, as it is modeled and discussed in the social sciences, is widely agreed to require time to propagate. Causes are generally assumed to precede their effects, and many social scientists, statisticians, and philosophers have claimed that time-ordering can be used for selecting appropriate causal models.

However, ordinary practice is typically neither explicit nor careful about the relationship between causation and time. In the social sciences, "time-free" causal models – i.e., ones which lack explicit information about the relative timing of causes and effects – are quite commonplace.[1] One might hope that the time-free models proposed in practice will always (or for the most part) be consistent with the assumption that causes precede their effects. But as we will see in this chapter, things are not so tidy. Often time-free causal models turn out to veil implausible temporal claims when examined closely.

Our goal in this chapter is to criticize this common practice of using time-free causal models. Our critique is not equivalent to a critique of a specific causal modeling methodology: it applies to regression models, structural equation models, and causal Bayes nets alike. Some statistical methodologies are better than others for causal modeling (for example, regression modeling is typically a poor choice for causal modeling, as has been shown by Glymour et al. (1994)). But our critique stands independent of these methodological questions and focuses on the widely affirmed maxim that causes precede their effects in time. If such a maxim is taken seriously, causal models that contain causes that *do not* precede their effects in time should be regarded as suspicious. We will show, with examples, that time-free causal models are able

---

[1] Economics is, perhaps, an exception. In economics, there has traditionally been more explicit attention to the temporal indices of variables. This is partly due to the nature of economic data, but also to a line of theoreticians who have given thought to the connection between causation and time. See Fisher (1970), Granger (1969), Hoover (2001).

to get away with violating the temporal ordering maxim simply because they do not exhibit their temporal claims on their face.

Here is how we will proceed. In Section 9.1, we examine an assertion commonly made by social scientists and methodologists – i.e., that causes precede their effects in time or that such an assumption aids causal search. We then point out that in the context of social science research, it is non-obvious what it means for a cause to precede its effects, and we provide a precise formulation of that platitude. In Section 9.2, we give an account of what is involved in "temporalizing" a time-free causal model – i.e., what is minimally required in order to explicitly state the temporal commitments of a causal model. We then formulate a minimal criterion according to which a time-free model can be judged "admissible" with respect to its possible temporalizations. In Section 9.3, we turn to practice. We illustrate how our criterion, despite its minimal requirements, places non-trivial constraints on the practice of causal modeling. We examine three cases from the social science literature that are problematic with respect to our criterion: Simons et al. (2009) on the influence of religiosity on risky sexual behavior, Timberlake and Williams (1984) on the influence of foreign investment on political exclusion in developing countries, and Beilock et al. (2010) on the influence of teacher math anxiety on student math achievement.

9.1 Temporal Ordering

Causal modelers and theorists in a variety of fields consistently assert that causes must be temporally prior to their effects. For example, in the first edition of his widely read introduction to structural equation modeling, psychologist and statistician Rex Kline (2011, 98) specifies a "temporal precedence" condition: "the presumed cause (e.g., X) must occur before the presumed effect (e.g., Y)." Similar assertions about the temporal priority of causes to effects abound elsewhere in commentaries about causal modeling across a wide range of social sciences.[2] It is further claimed that information about the temporal ordering of events can prove helpful in eliminating certain causal models and otherwise constraining causal search.[3]

---

[2] Compare statements by American political scientist Warren Miller (1999, 121), sociologist James Davis (1985, 11), statistician Paul Holland (1986, 946), and sociologist and statistician Michael Sobel (1990, 498). It should be mentioned that in econometrics, one sometimes sees so-called "simultaneous" equations, but as Fisher (1970) and Strotz and Wold (1960) have claimed, "simultaneous" is a misnomer in such models.

[3] For examples of such claims, see computer scientist Judea Pearl (2009, 42–43) and philosophers Spirtes et al. (2000, 93) and Glymour (2001, 41).

From such comments one might reasonably imagine that there is some well-defined "Causal Temporality Principle" that causal modelers across the social sciences collectively uphold. But defining such a principle is more difficult than might be expected given the widespread agreement. Let's make a first attempt at stating such a principle:

> NCTP   For any two variables X and Y, if X occurs at time t and Y occurs at time t' and X is a cause of Y, then $t < t'$.

NCTP here stands for *Naïve Causal Temporality Principle*. This principle might, at first glance, seem quite straightforward, especially to those steeped in a philosophical tradition that takes events to be the relata of the causal relation.[4] But it won't do. In the context of the social sciences, the X and Y appearing in NCTP are not events. They are *variables*.

One way in which the NCTP is naïve is that it treats the claim that a variable "occurs" as unproblematic. However, what it means for a variable to occur is not at all clear. After all, a variable is a measurable function over a set of statistical units. In other words, a variable is a set of ordered pairs, where each ordered pair consists of a unit and a value for that unit. So in order to make sense of the claim that a variable occurs, we would need to make sense of the claim that a set of ordered pairs occurs. But making sense of the claim that a set occurs is likely to be difficult for at least two reasons. First, a set is an object, and objects are not the sorts of things that occur. (An explosion, which is an event, occurs; an explosive does not.) Second, a set is an *abstract* object, and abstract objects present their own special challenges. The claim that an abstract object has any spatiotemporal properties at all is controversial.[5]

Alternatively, one might think of a causal variable as representing a property universal, such as height or color, which may be instantiated in different ways by different individuals. Property instances – tropes, such as a specific height or a specific color, rather than the individuals instantiating the properties – are then represented by evaluated variables. But even if

---

[4] See, e.g., Davidson (1967) or Lewis (1973a).
[5] One might argue that a set is grounded in its members and that if the members of a set are objects with spatiotemporal locations, then the set has a spatiotemporal location as well. But such a view is controversial, and the detailed story in the case of variables will be quite convoluted. See Rosen (2014), especially section 3, for an introduction to some of the issues. See Cook (2012) and Effingham (2012) for an exchange on whether sets are spatially located. See Caplan and Matheson (2004) and Korman (2014) for discussions of the relationship between abstract objects and time.

those individuals are temporally locatable, we still have not made it clear what it could mean for property universals to occur or to stand in temporal relations.

So let's try another approach. Instead of talking about the times at which some variables occur, we might try talking about the times at which some variables are measured. Spirtes et al. (2000, 93) make precisely this suggestion – i.e., that variables can be (partially) causally ordered on the basis of the time that measurements of the variables were taken. Focusing on measurement, we obtain the following *Revised Causal Temporality Principle*:

RCTP   Let X and Y be two arbitrary random variables, and let t and t' indicate the respective times at which the properties represented by X and Y are measured for the same unit u. If X is a cause of Y, then $t < t'$.

The revised principle is an improvement insofar as it gets rid of the language about variables "occurring" at particular times and instead talks about times at which the properties represented by the variables are measured. But applying the revised principle can lead to errors.

Consider the following scenario: suppose that a researcher is interested in studying the causal relationship between two variables A and B. For all of the units in her population, she measures the value of A at time t_A and the value of B at a later time t_B. And suppose she finds an association between A and B, and that she has independent reasons to think that the association is not explained by a common cause. Is she licensed to infer from the times at which they were measured that A is a cause of B? No. When we are dealing with variables whose values change over time, there is no guarantee that the temporal ordering of measurements will coincide with the temporal ordering of causal influence. Consider if B had actually been a cause of A, with the following data-generating mechanism and time traces as shown in Figure 9.1:

$$A_t = 0.9 \cdot A_{t-1} + 0.1 \cdot B_{t-1} + \epsilon_A$$
$$B_t = B_{t-1} + \epsilon_B$$

When we are dealing with variables (or the properties variables represent), there is no guarantee that the temporal ordering of our measurements will coincide with the temporal ordering of the causal influences we care about. It is

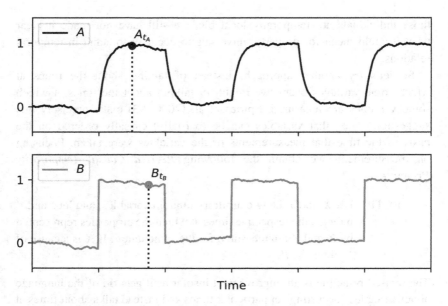

Figure 9.1 In the time traces above, changes in B trigger a more gradual change
in A. The time order of measurements (A measured at t_A and B measured at t_B,
where $t_A < t_B$) cannot be taken to provide information about the true causal
ordering of A and B.

a mistake to draw an inference from the temporal ordering of the measurements
to the causal ordering of the variables being measured. The temporal ordering
that interests us – i.e., the temporal ordering of causal influence – is independent of when the variables were measured. A measurement of a variable at a
particular point in time gives us an estimate of the variable's value at that point
in time, but it does not inform us as to when the variable acquired the value in
question.

What we really want to get at is the question of how properties change
their values over time in relation to one another. If, for example, we have
two associated variables A and B with no common cause, and we know that
the moment at which the value of the underlying property a changes always
occurs before the moment at which the value of the underlying property b
changes, then we would conclude that A is the cause of B rather than vice
versa. Conversely, if we know that C is a cause of D, we would presume
that the changes in the value of the underlying property d will always be
preceded in time by any relevant changes in the value of the underlying
property c.

Hence, we want a principle like the following:

CTP   For any two random variables X_t and $Y_{t'}$, where the indices t and t' indicate the times at which properties x and y take on their respective values, if X_t is a cause of $Y_{t'}$, then $t < t'$.[6,7]

We think that the above version of the Causal Temporality Principle (CTP) captures fairly well what is meant by many of the platitudes in the literature about causes preceding their effects, and so it is the one we will use throughout the remainder of this chapter. It should be immediately clear, however, that there will be a significant and unavoidable epistemic problem for causal modelers when the exact order in which properties change their values remains unknown. For this reason and other reasons we will explore below, the temporal ordering of causal variables will be a much messier affair in practice than one might expect, even assuming the CTP.

9.2 Temporalization

We must now confront a problem. As we mentioned at the outset of this chapter, "time-free" causal models – i.e. ones which lack explicit information about the relative timing of causes and effects—are quite commonplace in actual practice. But our CTP makes use of time-indexed variables. How then are we to use CTP as an assessment tool for models which lack time indices? The answer is that when assessing a time-free model, we must consider its possible *temporalizations*.

---

[6] One might argue that even this version of the CTP is not adequate, for variables might "feed back" on each other over time as opposed to taking on their values suddenly at one moment. The issue of causal feedback loops and the way that multiple variables might causally interact during gradual value changes is a complex one that we will not attempt to address in this chapter; rather, we focus on examples that involve much simpler temporal problems.

[7] Whether the CTP is generally true, and if so, whether it is *necessarily* true or only contingently true is unclear. Some philosophers – e.g., Ben-Yami (2007, 2010), Black (1956), Flew in Dummett and Flew (1954), Flew (1956, 1957), Gale (1965), and Mellor (1991) – have argued that causes precede their effects necessarily. Other philosophers – e.g., Dummett in Dummett and Flew (1954), Dummett (1964), Earman (1976), Garrett (2014, 2015), Reichenbach (1956), Roache (2009), and Scriven (1956) – have argued that causes precede their effects only contingently, if they do so at all. In contemporary physics, the relationship between causation and time is still up for grabs. Causal influence might be instantaneous or even run backwards sometimes, if certain physical theories turn out to be true – see Carroll (2010), Earman et al. (2009), and Smeenk and Wüthrich (2011). The fact that instantaneous and even backwards causation is taken seriously in contemporary physics not only suggests that the CTP is at best only contingently true; it also suggests that the CTP might be true only for some restricted range of cases, e.g., middle-sized dry goods. One virtue of having an explicit principle relating causation and time-ordering, however, is that the principle may be adopted or rejected depending on the context.

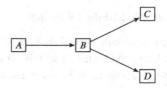

Figure 9.2 Simple time-free causal graph.

At a minimum, temporalizing a causal model involves assigning a partial ordering to the variables in the model in order to represent their temporal relations. In order to see how temporalization works, consider the simple time-free graph in Figure 9.2.

A minimally informative temporalization of the graph in Figure 9.2 could be provided by first pairing a time-index variable with each causal variable and then assigning a partial-ordering to the set of time-index variables. For example, we could temporalize the graph by stipulating the following:

1 A takes on its value at time t_A.
2 B takes on its value at time t_B.
3 C takes on its value at time t_C.
4 D takes on its value at time t_D.
5 $t_A < t_B < \{t_C, t_D\}$.

We will consider a temporalization of this sort when we discuss a cross-sectional case study in Section 9.3.1. Alternatively, a slightly more informative temporalization might provide information about the "lag times" between a cause variable taking on its value and each of its effects doing so:

1 A takes on its value at time t_A.
2 B takes on its value at time t_B, which is one year later than t_A.
3 C takes on its value at time t_C, which is one year later than t_B.
4 D takes on its value at time t_D, which is one year later than t_B.

The level of precision for temporalizing a causal model is relative to a choice of time points, which may be more or less finely grained. For example, suppose one takes the relevant time points to be years. A temporalization relative to such a choice assigns a year to each variable in the model. (In Section 9.3.2 we examine a case study in which time points like these are given, although, as we will see, some are actually aggregate measurements over a period of time.) Here is one way such a temporalization might go:

1 A takes on its value in the year 2016.
2 B takes on its value in the year 2017.
3 C takes on its value in the year 2018.
4 D takes on its value in the year 2018.

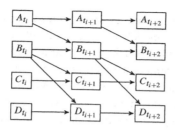

Figure 9.3 A time-slice temporalization of Figure 9.2, where it is stipulated that $t_i < t_{i+1} < t_{i+2}$.

Sometimes there are cases when it is suspected that one or more of the variables in question change their values over time in a way that necessitates a more temporally fine-grained representation of the causal relationships. In such cases, temporalization of a time-free graph can involve splitting causal variables into *sets* of variables that take on their respective values at distinct time points. We call this type of temporalization a *time-slice temporalization*. For example, the causal graph pictured in Figure 9.3 is one possible time-slice temporalization of the causal graph in Figure 9.2. Time-slice temporalization is especially useful in cases of dynamic causal systems that involve causal feedback loops. (We see an example of time-slice temporalization in Section 9.3.3)

At this point, two observations about the temporalization of time-free causal graphs are in order. First, a time-free causal model may be temporalized in an infinite number of ways: there are no in-principle constraints on the temporalization of a time-free graph. A minimal temporalization of the causal graph in Figure 9.2 could very well stipulate, for example, that $t_D < t_C < t_B < t_A$, which would amount to the claim that the causal relations of the model all flow backward in time. Because temporalization itself is not constrained, only a subset of the possible temporalized versions of a given time-free causal model will satisfy the CTP.[8]

Second, since time-free causal models have infinitely many possible temporalized versions, time-free causal models *underdetermine* their temporalizations.[9] Temporal underdetermination can make the results of important policy

---

[8] In this way, our project differs from that of Spirtes et al. (2000), who attempt to characterize the precise circumstances under which a time series graph such as the one in Figure 9.3 can be collapsed into a time-free graph. Our aim is to leave temporalization a completely unconstrained and underdetermined question; in constrast, Spirtes examines criteria for assuring that a translation between time-free and time-indexed graphs is legitimate.

[9] Talking about different "versions" of the same causal model raises many interesting philosophical challenges. A careful treatment of statements such as "Causal model M_1 is a temporalized version of causal model M_2" would be valuable, but such a treatment is beyond the scope of the present chapter.

questions difficult (and sometimes impossible) to answer, and the practical value of time-free causal models is correspondingly limited. Temporalization increases the practical utility of a time-free causal model, and more informative temporalizations (e.g., temporalizations that provide information about lag times or that assign time points to variables) will be of greater practical use than minimally informative temporalizations (i.e., temporalizations which only provide a partial ordering over the time indices of the variables).

We are now in a position to formulate a plausible criterion for admissibility of a time-free causal graph, if it is to obey the widely accepted requirement that causes precede their effects:

TEMPORALIZATION CRITERION OF ADMISSIBILITY: A time-free causal model is *admissible* if and only if there is at least one temporalized version of the model that satisfies the CTP while also being responsible to the data.

To accept the above criterion is to make a relatively minimal assumption about time-free causal models: that for any time-free causal model, there is some way of interpreting it such that all of its causes precede their respective effects in time, and that the data from which the model is inferred are consistent with such a temporal ordering. In other words, one might expect that when a social scientist proposes a time-free causal model, he or she has in mind a plausible temporal ordering that is consistent with the data and which could be supplied on demand.

One might suppose that the temporalization criterion of admissibility will be trivial to satisfy in virtue of the fact that time-free causal models so radically underdetermine how they are to be temporalized. But surprisingly, time-free causal models cannot always be temporalized in a way that satisfies the CTP while *also* satisfying reasonable and justifiable assumptions about the temporal properties of the variables actually measured *and* appropriately representing the statistical facts. In other words, time-free causal models cannot always be temporalized in a way that satisfies the CTP while also being responsible to the data. We give three examples of such failures in Section 9.3.

9.3 Responsible Modeling

As we have noted above, every time-free causal model has many possible temporalizations. On that basis, it seems reasonable to expect that if a time-free causal model has been proposed by a social scientist on the basis of empirical data, at least one of its temporalizations would be both plausible with respect to its temporal ordering and also responsible to the empirical data on which the time-free model was based. Our criterion of admissibility is founded on precisely this expectation. But we will show in this section, by considering three examples of causal modeling in the social sciences, that these minimal expectations are not always satisfied.

Here we discuss three studies: Simons et al. (2009) on the influence of religiosity on risky sexual behavior, Timberlake and Williams (1984) on the influence of foreign investment on political exclusion in developing countries, and Beilock et al. (2010) on the influence of teacher math anxiety on student math achievement. The time-free models we consider here initially appear to be reasonable. But when one thinks carefully about how each of the models might be temporalized, one quickly realizes that either plausible temporal ordering or responsibility to the data must be sacrificed.

Thus, each of the cases we consider below fails to satisfy our temporalization criterion of admissibility. In each case, we encounter a temporalisation dilemma: we can *either* temporalize the time-free model in a way that satisfies the CTP *or* in a way that is responsible to the data, *but we cannot do both.*

Roughly speaking, a temporalization that is responsible to the data must assign time indices to variables in a manner that is consistent with the temporal features of the data as they were collected or generated. What will it look like for a model to *fail* to be responsible to the data? We will discuss a variety of failures in this section. We will see some failures that involve inadequate study design (e.g., questionnaires that result in temporally ambiguous data). We will see other failures that are problems with data generation and collection (e.g., unsophisticated use of time-aggregate variables). We will also see a case in which, in the process of temporalizing a causal model in a manner consistent with the CTP, we end up with a temporalized graph that fails to accurately represent the statistical relationships observed in the data.

Despite the variety of specific failures we will see in this section, a unifying thread runs through all of the cases we will encounter here. In each case, very little attention is paid to the timing and temporal ordering of the variables being used in the model. As a result, each case study produces a time-free causal model that appears reasonable at first glance, but on closer inspection, fails to have an empirically justified temporalization that is consistent with the CTP.

9.3.1 Religiosity and Sexual Behavior

Simons et al. (2009) investigated the influence of religiosity on sexual behaviors among adolescents. They used pencil and paper surveys to collect data

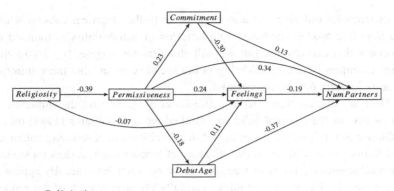

$$Religiosity = \epsilon_R$$
$$Permissiveness = -0.39 \cdot Religiosity + \epsilon_P$$
$$Commitment = 0.23 \cdot Permissiveness + \epsilon_C$$
$$DebutAge = -0.18 \cdot Permissiveness + \epsilon_D$$
$$Feelings = -0.07 \cdot Religiosity - 0.30 \cdot Commitment$$
$$+ 0.24 \cdot Permissiveness + 0.11 \cdot DebutAge + \epsilon_F$$
$$NumPartners = 0.13 \cdot Commitment - 0.19 \cdot Feelings$$
$$+ 0.34 \cdot Permissiveness - 0.37 \cdot DebutAge + \epsilon_N$$

Figure 9.4 Graphical model and structural equation model for females.

from 2,108 undergraduates in sociology courses (enrolled in the academic year 2001–2002). For each student, Simons et al. measured the following variables: *Religiosity* (influence of religion on daily life), *Permissiveness* (attitude about sexual permissiveness), *Commitment* (level of commitment to first sex partner), *Feelings* (feelings about first experience with intercourse), *Debut Age* (age at time of first experience with intercourse), and *NumPartners* (number of partners with whom the individual has had sexual intercourse).

Simons et al. used the statistical software AMOS 5.0 to construct a linear structural equation model for their data. They found a significant difference in the models for females and males, which are shown in Figures 9.4 and 9.5, respectively.[10] Simons et al. interpret their models causally and take them to have important policy implications insofar as they characterize the mechanisms by which religiosity influences various aspects of sexual behavior. Granting the causal interpretation, how should we understand the temporal relationships between the variables in the models?

[10] For purposes of presentation, we have suppressed the error terms in the graphical models. Also, as a technical aside: in each model, the error terms for commitment and sexual debut are correlated; hence, the models are not recursive, even though the causal graphs are acyclic.

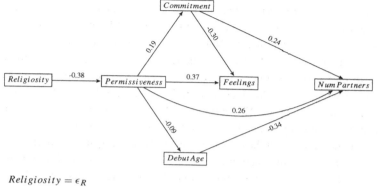

$$Religiosity = \epsilon_R$$
$$Permissiveness = -0.38 \cdot Religiosity + \epsilon_P$$
$$Commitment = 0.19 \cdot Permissiveness + \epsilon_C$$
$$Debut Age = -0.09 \cdot Permissiveness + \epsilon_D$$
$$Feelings = -0.30 \cdot Commitment + 0.37 \cdot Permissiveness + \epsilon_F$$
$$Num Partners = 0.24 \cdot Commitment + 0.26 \cdot Permissiveness - 0.34 \cdot Debut Age + \epsilon_N$$

Figure 9.5 Graphical model and structural equation model for males.

Consider the simpler model for males (Figure 9.5). If we adhere to the CTP, the model entails the following relations between the temporal indices of the variables:

(1) Since *Permissiveness* is an effect of *Religiosity*,

$t_{Religiosity} < t_{Permissiveness}$;

(2) Since *Commitment* and *Debut Age* are both effects of *Permissiveness*,

$t_{Permissiveness} < t_{Commitment}$ and $t_{Permissiveness} < t_{Debut Age}$, but $t_{Commitment}$ and $t_{Debut Age}$ have no definite temporal ordering between themselves.

(3) Since *Feelings* and *Num Partners* are both effects of *Commitment*, while *Num Partners* alone is an effect of *Debut Age*, then using (2),

$t_{Commitment} < t_{Feelings}$, $t_{Commitment} < t_{Num Partners}$, $t_{Debut Age} < t_{Num Partners}$, but $t_{Feelings}$ and $t_{Num Partners}$ have no definite temporal ordering between themselves.

(4) Therefore, $t_{Religiosity} < t_{Permissiveness} < \{t_{Commitment}, t_{Debut Age}\} < \{t_{Feelings}, t_{Num Partners}\}$.

However, the above relations between temporal indices do not make much sense in the context of the empirical study as it was performed. Consider the value of *Religiosity* and its supposed temporal precedence to all others, as reflected in claim (4) above. In the study, the value of *Religiosity* was assessed using a single question on the survey: "What is the influence of religion in your daily life?"[11] Notice that the question makes no reference to how much the individual was influenced by religion at a prior time (e.g., prior to any sexual activity), but rather appears to ask a question about the extent to which the individual feels influenced by religion at the time the survey is being filled out. The variable *Religiosity*, therefore, most plausibly measures something about the respondent's life at the moment at which he filled out the survey. But the causal model constructed by the researchers places *Religiosity* as the first variable in several causal chains. Such a model, if it is to satisfy the CTP, would require that the measured value of *Religiosity* reflect the value of the variable at a point in time earlier than when all other variables took on their values (as claimed in (4) above). The survey instrument does not plausibly reflect a measurement of the value of *Religiosity* at such an early point in time. The value of *Religiosity* can change a great deal during the adolescent years, and hence, there is no guarantee that the value of *Religiosity* measured by the survey is equal to the value it had at the early adolescent time point of interest.

Now, consider the variable *Permissiveness*. According to claim (4), *Permissiveness* takes on its value after *Religiosity* and before all other variables. Simons et al. measured this variable by using the Sexual Permissiveness Scale created by Reiss (1967) in which the individual is asked to indicate on a five-point scale how strongly he or she agrees that sexual intercourse is acceptable in different types of relational situations (ranging from a first date to marriage). Again, it does not appear that the survey administered by Simons et al. gave any indication that the individual was to give the opinions about sexual permissiveness that he had held at an earlier point in time rather than those he presently held. The lack of temporal clarity in the question makes it difficult to interpret the data. How are we to know the point in time about which a given respondent is reporting? Since an individual's views about sexual permissiveness may easily change over time, the measured value of *Permissiveness* may not match the value of the variable at the time point that is of interest in the model.

Finally, there was temporal ambiguity in the survey question about the individual's first sex partner, which asked, "With whom did you first have sexual intercourse?" The response choices for this question were: "have never had intercourse," "spouse after marriage," "fiancé," "steady dating partner," or

---

[11] Students were asked to respond on a numerical scale from 1, anchored at "none," to 5, anchored at "very influential."

"casual acquaintance." In order for the temporal ordering of the study variables to be clear, it would be important for the survey respondents to characterize their first sexual partner according to the nature of the relationship at the time of first intercourse rather than to report about the current state of that relationship, given that the nature of the relationship might have later changed. For example, if two casual acquaintances have sex and later begin to date steadily, responses of "steady dating partner" or "casual acquaintance" might both be reasonable.[12] There is no indication in the published account that care was taken on this point, so there is no assurance that the measured value of *Commitment* is an accurate report of the nature of the relationship with that partner at the time point of interest in the model. By (4), *Commitment* must take on its value after *Religiosity* and *Permissiveness* but before *Feelings* and *NumPartners*.

We have analogous worries about the temporal ordering of the variables in the model for females (Figure 9.4), but they need not be rehearsed here. In both cases, the temporal ambiguity in the data leads to a Temporalization Dilemma. On the one hand, we could presume the temporal ordering of causes and effects that would follow from the time-free model by the CTP. But in that case the temporal ordering is not well-justified by the temporally-ambiguous survey data, and hence, the temporalized model would not have adequate empirical support. On the other hand, we could accept the causal model as the most empirically adequate one given the available data. However, since the data were not measured in a way that reflects the temporal ordering prescribed by the CTP, being responsible to the data would imply rejecting the CTP and with it the common-sense ordering of cause and effect. For if we accept the model's claim that, e.g., *Religiosity* is an indirect cause of *DebutAge* along with the most plausible time-ordering of the survey data, then we are admitting a causal relationship that points backward in time.

Summing up: Because Simons et al. were not careful or explicit about the temporal relations among their variables, the models they propose are not admissible according to our criterion. Consequently, they face a Temporalization Dilemma: Either (a) reject common-sense temporalizations of the causal model because these temporalizations are not reflected by the manner in which the data were measured, or (b) accept the causal model and reject the common-sense temporal ordering of cause and effect.

---

[12] It is difficult to say what to do with the answer "spouse after marriage," since that value includes some temporal information. And it is odd that something similar was not done for the other values: e.g., "fiancé after engagement."

9.3.2 Foreign Investment, Political Exclusion, and Repression in Third World Nations

Our second example – one that has been much discussed in the literature on causal inference – is the examination by Timberlake and Williams (1984) of the impact of foreign investment on political exclusion and repression in third-world nations.[13] Timberlake and Williams hypothesized that increased investment in third-world nations by more developed nations would increase the number of citizens excluded from participation in government and decrease the civil liberties of the citizenry at large. They write:

> The elite who are enriched by the penetration of foreign capital have an interest in assuring the security and profitability of that capital. Important ways these interests are pursued include promoting political structures which exclude opposition to elite interests and supporting repressive regimes which will impose negative sanctions when opposition is organized. [...] We are led, then, to the hypothesis that ties of dependence contribute directly to political exclusion and to government repression. (142)

In order to test their hypotheses, Timberlake and Williams studied a sample of 72 countries determined to be of peripheral and semiperipheral status on the international stage. Government repression was measured by looking at the negative sanctions imposed on citizens by government authorities; they aggregated data over the five-year period of 1973–1977 to produce the variable $SAN77$ (where "SAN" stands for "sanctions"). Conflict in opposition to the state was measured by looking at the number of protest demonstrations carried out against the state, one of its leaders, its policies, or its actions. These data were likewise aggregated over the same five-year period to produce the variable $PRO77$ (where "PRO" stands for "protests").

Their initial analysis made use of five other variables. Political exclusion ($POLEX$) was measured using Gastil's "Political Rights Index," which is given on a seven-point scale from fewest people excluded (1) to most people excluded (7). Civil liberties ($CIVIL$) were measured by the "Civil Rights Index," also due to Gastil and given on a seven-point scale, from extensive civil liberties (1) to very restricted civil liberties (7). Both of these variables

---

[13] The article is discussed in Cartwright (1989), Freedman (1997, 2009), Glymour (1999), Glymour et al. (1987), Spirtes et al. (2000), and Spirtes and Scheines (1997). We are largely in agreement with the criticisms leveled by these writers and with the positive approach to causal search advocated by Spirtes et al. (2000). However, we think that there is more to be said about the Timberlake and Williams example. In thinking about temporalization, we have been struck by the thought that the Timberlake and Williams model is broken before the mistakes they make in their statistical analysis. Before reaching the sorts of problems pointed out by Freedman (2009) or, more particularly, by Spirtes et al. (2000), the Timberlake and Williams model is already in serious trouble.

are temporally ambiguous; the data source used by Timberlake and Williams contains both of Gastil's indices for the years 1973–1979, but Timberlake and Williams do not report how their variables $POLEX$ and $CIVIL$ relate to Gastil's numbers over those seven years. $POLEX$ and $CIVIL$ might be snapshots from a particular year, but since Timberlake and Williams do not indicate a particular year, we think it's likely that they used an aggregate over the seven years from 1973–1979. Energy consumption ($ENERGY$) was measured in coal equivalents for the year 1975, and population (POP) was also measured for the year 1975. Economic dependence, or "penetration by foreign investment" ($PEN73$), was given by a transformed measure of the dollar value for stock of foreign investments in 1973 that controlled for the size of the penetrated country's economy.[14]

The hypothesis that foreign investment increases political exclusion was tested by regressing $POLEX$ on $PEN73$, $CIVIL$, and $ENERGY$. Timberlake and Williams thus generated the following regression equation:

$$E(POLEX|PEN73, ENERGY, CIVIL) =$$
$$0.743 + \mathbf{0.762} \cdot PEN73 - \mathbf{0.478} \cdot ENERGY + \mathbf{1.061} \cdot CIVIL,$$

which, if interpreted causally, suggests that, indeed, increasing foreign investment increases political exclusion.[15,16] This model is represented graphically in Figure 9.6.

To test the hypothesis that foreign investment causes government repression, Timberlake and Williams used two-stage least squares (2SLS) regression to estimate the coefficients on a pair of simultaneous equations for $SAN77$ and $PRO77$. The model they constructed shows mutual causation between government sanctions and protests. In equation form the model is:

---

[14] Timberlake and Williams used the third edition of the *World Handbook of Political and Social Indicators* (Taylor and Jodice, 1983) as a data source for all of their variables except $PEN73$. The *book* contains both of Gastil's indices (the Political Rights Index and the Civil Rights Index) for the years 1973–1979. Timberlake and Williams also refer at several points to Gastil (1978), which contains his numbers for 1978, but they only reference it in relation to his methodology.

[15] Throughout their paper, Timberlake and Williams move quite fluidly between claims about associations among variables and claims about the effects of certain variables on certain others. It would be disingenuous for them to deny that they intended for their regression equations to be interpreted causally.

[16] Spirtes et al. (2000) objected to the causal interpretation of Timberlake and Williams' regression model. Instead of ordinary regression, they considered the output of the PC and FCI algorithms given the same data. On the basis of an equivalence class of causal models, they claimed that foreign investment penetration does not cause political exclusion, contrary to the conclusion of Timberlake and Williams. But they did not address the bizarre temporal features of the data any more than did Timberlake and Williams.

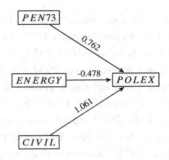

Figure 9.6 Graphical model for causes of $POLEX$.

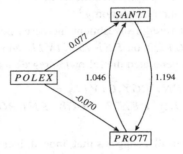

Figure 9.7 Graphical model for causes of $SAN77$ and $PRO77$.

$$E(SAN77|PRO77, POLEX, PEN73, POP) =$$
$$-0.130 + \mathbf{1.046} \cdot PRO77 + \mathbf{0.077} \cdot POLEX$$
$$+0.059 \cdot PEN73 + 0.025 \cdot POP,$$
$$E(PRO77|SAN77, POLEX, ENERGY, POP) =$$
$$0.473 + \mathbf{1.194} \cdot SAN77 - \mathbf{0.070} \cdot POLEX$$
$$-0.114 \cdot ENERGY - 0.122 \cdot POP.$$

This 2SLS model is represented graphically in Figure 9.7.[17]

By connecting this model to the one constructed using ordinary regression, Timberlake and Williams hoped to determine whether foreign investment has an indirect effect on government sanctions through political exclusion. In this way, they hoped to learn something about the mechanism by which

---

[17] In their analysis, Timberlake and Williams consistently make causal claims in those cases where the coefficients are statistically significant at the 0.05 level and reject causal claims where they are not. Those that are statistically different from zero at the 0.05 level we have written in bold face. We have also depicted the arrows in our causal graphs accordingly.

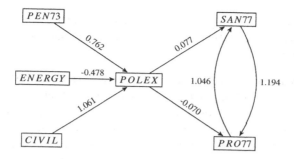

Figure 9.8 Combined graphical model.

foreign investment increases government repression (assuming it does so). If we combine the graphs from both models, we get the graph in Figure 9.8.

Now, if we combine the causal claims represented in the graph in Figure 9.8 with the temporal features of the data, we get the following:

(1) *PEN73* (measured at $t = 1973$) is a cause of *POLEX* (an aggregate of measurements made from 1973–1979).[18]

(2) *ENERGY* (measured at $t = 1975$) is a cause of *POLEX* (an aggregate of measurements made from 1973–1979).

(3) *CIVIL* (an aggregate of measurements made from 1973–1979) is a cause of *POLEX* (an aggregate of measurements made from 1973–1979).

(4) *POLEX* (an aggregate of measurements made from 1973–1979) is a cause of *SAN77* (an aggregate of measurements made from 1973–1977).

(5) *POLEX* (an aggregate of measurements made from 1973–1979) is a cause of *PRO77* (an aggregate of measurements made from 1973–1977).

(6) *PRO77* (an aggregate of measurements made from 1973–1977) is a cause of *SAN77* (an aggregate of measurements made from 1973–1977).

---

[18] As noted above, the relationship between the variable *POLEX* and values of the Political Rights Index for the years 1973–1979 is not made clear in Timberlake and Williams' study. The lack of clarity on this point is problematic in itself and is a confirming instance of the main message of this chapter about the often careless practice of causal modeling with respect to time. Our best guess, as stated above, is that the variable *POLEX* is an aggregate of the Political Rights Index values over the years 1973–1979, and if so, our observations follow. The same goes for the variable *CIVIL*.

(7) $SAN77$ (an aggregate of measurements made from 1973–1977) is a cause of $PRO77$ (an aggregate of measurements made from 1973–1977).

Clearly, many of the causal claims being made here are temporally peculiar. For example, claim (2) states that a variable representing a fact about 1975 is a cause of a variable that is an aggregate over a span of years that includes years prior to 1975. Similarly, claims (4) and (5) state that a variable aggregated over the years 1973–1979 is a cause of a variable aggregated over the years 1973–1977. Such claims seem to violate CTP, since they seem to require that variables indexed by later years affect variables indexed by earlier years.

What should we make of causal claims involving variables aggregated over periods of time? The CTP gives us no guidance on the temporal ordering of such aggregates, but is there some other charitable reading of such cases? We might imagine one: Susan's yearly income might be said to have an influence on her yearly expenses over the same year, even if— especially if—the only measures we have of her income and expenses are aggregates over the year. Might not causal relationships among aggregate variables be a kind of summary statement about influences over smaller periods of time?[19] Perhaps. However, we think it is incumbent on researchers to provide an explanation and interpretation of causal claims that involve variables with idiosyncratic temporal features. Timberlake and Williams do not acknowledge that there is anything unusual in their causal claims. In this particular case, it appears that the variables were chosen as a matter of convenience – i.e., they were the numbers that were available – but such a practical justification is not enough. An argument needs to be made for how data with a particular structure justifies the kind of inferences one wants to make.

The temporal peculiarity of Timberlake and Williams' analysis does not end there. According to the model depicted in Figures 9.7 and 9.8, the variables $SAN77$ and $PRO77$ are mutual causes. This finding led Timberlake and Williams to construct a new model including the lagged measures $SAN72$ and $PRO72$ (which they formed using aggregates over the five-year period 1968–1972) for sanctions and protests. In equation form, the new model is:

---

[19] We thank an anonymous reviewer for suggesting this example of a charitable interpretation of causal claims among aggregate variables.

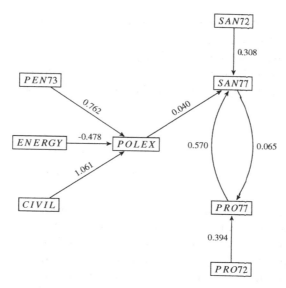

Figure 9.9 Combined graphical model with lagged versions of PRO and SAN added.

$E(SAN77|PRO77, POLEX, PEN73, ENERGY, POP, SAN72) =$
$-0.412 + \mathbf{0.570} \cdot PRO77 + \mathbf{0.040} \cdot POLEX - 0.003 \cdot PEN73$
$+0.086 \cdot ENERGY + 0.091 \cdot POP + \mathbf{0.308} \cdot SAN72$

$E(PRO77|SAN77, POLEX, PEN73, ENERGY, POP, PRO72) =$
$0.217 + \mathbf{0.065} \cdot SAN77 - 0.045 \cdot POLEX - 0.051 \cdot PEN73$
$-0.022 \cdot ENERGY - 0.059 \cdot POP + \mathbf{0.394} \cdot PRO72$

This model, when combined with the first regression equation depicted in Figure 9.6, is represented graphically in Figure 9.9 (where again, those variables whose coefficients are not statistically significant are not considered causes).

Timberlake and Williams leave their choice of a five-year lag mysterious. Additionally, once they choose to introduce the time-lagged versions of the two variables SAN and PRO, it becomes an open question why other variables in the model should not have their own lagged versions. The result is a set of causal relations among variables whose temporal relationships seem to have been arbitrarily chosen. The arbitrary nature of the temporal relationships, although not apparent at first glance, is made evident when drawn out explicitly. In Figure 9.10, we have labeled the temporal properties of the causal relationship along each edge of the graph.

Let's step back and look at the interpretational problem we are grappling with in this example. If we choose to interpret Timberlake and Williams' final causal model (Figure 9.9) according to the CTP, we should expect the following:[20]

$$\{t_{PEN73}, t_{ENERGY}, t_{CIVIL}\} < t_{POLEX}$$
$$\{t_{POLEX}, t_{SAN72}\} < t_{SAN77}$$
$$t_{PRO72} < t_{PRO77}.$$

Alternatively, we might interpret the temporal relationships among the causal variables to be the same as the temporal relationships among the measurements and aggregates as they were carried out in the study. For example, we could simply interpret the graph as saying that the variable $POLEX$, the value of which is generated by an aggregation procedure over a seven-year period, causally influences the variable $SAN77$, the value of which is generated by an aggregation procedure over an overlapping but earlier five-year period. Such a claim would make the model responsible to the procedure by which the data were generated, but it would be a peculiar claim, since it is unclear what it means for a time-aggregate variable to cause another time-aggregate variable. The claim becomes even more peculiar in the face of the fact that the cause variable is an aggregate over a period that includes time points *later* than those included in the aggregated effect variable. So again, we encounter the Temporalization Dilemma: we have the choice of *either* an interpretation of the model that is responsible to the data *or* an interpretation of the model that is consistent with common-sense ideas about causal-temporal precedence. We do not see how to construct a fully temporalized version of Timberlake and Williams' final model that satisfies both desiderata.

9.3.3 Female Teachers' Math Anxiety and Girls' Math Achievement

Our third and final example comes from Beilock et al. (2010), who wondered what effect the math anxiety of elementary educators might have on their students. To find out, Beilock et al. made a number of observations on 17

---

[20] Note that even with the CTP, there is no consistent way of giving an ordering to t_{SAN77} and t_{PRO77} because they are mutual causes. The CTP gives no guidance for intepreting causal cycles temporally.

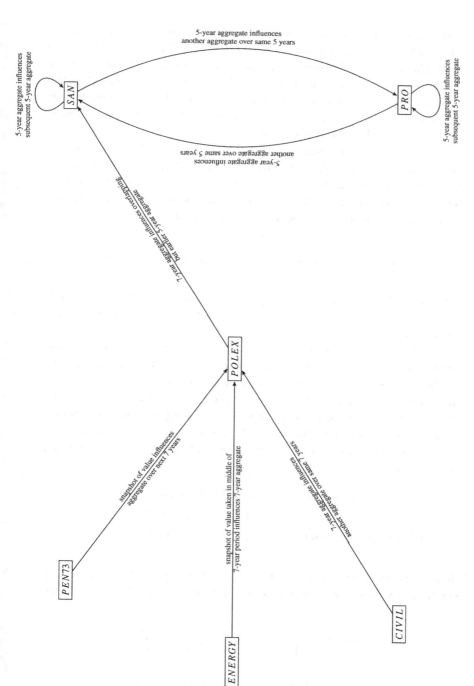

Figure 9.10 Timberlake and Williams' final causal model with temporal relationships labeled on edges.

first- and second-grade classes in one large Midwestern urban school district. They measured the math ability of the 117 students (both boys and girls) in those classes at two different points in the school year – in the first three months and again in the final two months. Beilock et al. also measured student gender ability beliefs at the beginning and end of the school year. Finally, they measured the math ability and the math anxiety of the 17 female teachers in the last two months of the school year.

Beilock et al. found no significant relation between a teacher's math anxiety and students' math achievement at the beginning of the school year. However, they report that "regression analysis established that teachers' math anxiety had a significant negative effect on girls' [but not boys'] math achievement at the end of the school year." That is, for the theoretical model

$$[Girls'\ Math\ Achievement] = \beta \cdot [Teacher\ Math\ Anxiety] + \epsilon,$$

they estimated a statistically significant negative coefficient, namely $\hat{\beta} = -0.21$. They drew the directed graph in Figure 9.11 to represent this causal claim.

However, they also found a significant association between a teacher's math anxiety and female students' gender ability beliefs at the end of the year. Furthermore, a female student's gender ability beliefs were associated with her math achievement at the end of the year. When Beilock et al. regressed girls' math achievement on both the teacher's math anxiety and the girls' beliefs about gender ability, they found that the gender ability beliefs alone were a significant predictor of math achievement level. That is, for the following theoretical model:

$$[Girls'\ Math\ Achievement] = \beta_1 \cdot [Teacher\ Math\ Anxiety]$$
$$+ \beta_2 \cdot [Gender\ Ability\ Beliefs] + \epsilon_1$$
$$[Gender\ Ability\ Beliefs] = \beta_3 \cdot [Teacher\ Math\ Anxiety] + \epsilon_2,$$

they found coefficient estimates of $\hat{\beta}_2 = -0.23$ and $\hat{\beta}_3 = 0.31$, where both coefficients were statistically significant, but found that coefficient β_1 was not

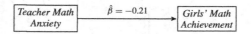

Figure 9.11 Beilock et al.'s simple graphical model.

statistically different from zero. So a teacher's math anxiety was associated with girls' math achievement, but that association ceased to be statistically significant if one conditioned on the gender ability beliefs of the female students. Beilock et al.'s causal conclusion based on this analysis was that a (female) teacher's anxiety about mathematics negatively affects the math ability of her female students by inducing a belief that girls are not good at math. Accordingly, Beilock et al. drew the causal graph in Figure 9.12.

Summarizing their results, Beilock et al. write: 'In early elementary school, where the teachers are almost all female, teachers' math anxiety carries consequences for girls' math achievement by influencing girls' beliefs about who is good at math' (1860).

When they draw the causal graph of Figure 9.12 on the basis of their regression analysis, we take Beilock et al. to be implicitly endorsing the claim that independence has a default status and the claim that a causal graph should entail all and only the conditional independence claims required by the data. If a statistical test fails to reject the claim that two variables A and B are independent conditional on some set $\{C_i\}$ of variables, then one ought to accept that A and B are, in fact, conditionally independent given $\{C_i\}$, and one ought not to draw an edge connecting A and B.[21]

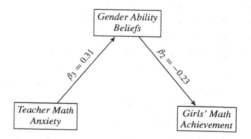

Figure 9.12 Beilock et al.'s graphical model including *Gender Ability Beliefs* as a mediator.

---

[21] In criticizing Beilock et al.'s model, we use the standard for responsibility to the data that we have just articulated and take them to endorse. However, as Conor Mayo-Wilson reminded us, one might be suspicious of these commitments. Specifically, the move from (1) the failure to observe a statistical dependence between *Teacher Math Anxiety* and *Girls' Math Achievement* conditional on *Gender Ability Beliefs* to (2) the claim that there is no direct causal influence from *Teacher Math Anxiety* to *Girls' Math Achievement* is potentially problematic. Mayo-Wilson suggested that Beilock et al. might repair their model by picking the causal graph that adds the fewest dependencies to those required by the data. We are not sure whether the suggested strategy would work as a general rule for bringing models into line with the Temporalization Criterion. But in any event, the model recommended by Mayo-Wilson's strategy is not the model Beilock et al. actually produced. The Beilock et al. model remains inadmissible according to our criterion (on what we take to be Beilock et al.'s standard for

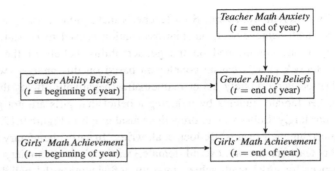

Figure 9.13 Temporalized version of Beilock et al.'s graphical model including *Gender Ability Beliefs* as a mediator.

The causal models pictured in Figures 9.11 and 9.12 do not have explicit temporal information. But the measurements Beilock et al. made were temporally ordered. Beilock et al. measured students' gender ability beliefs and math ability twice – once at the beginning of the school year and once at the end. (*Teacher Math Anxiety* was measured once at the end of the school year.) Thus, the temporal indices are relatively clear in this case. If we assume that gender ability beliefs at the beginning of the year affect gender ability beliefs at the end of the year and if we assume that math ability at the beginning of the year affects math ability at the end of the year, then the causal graph in Figure 9.13 seems reasonable – at least at first blush.

The graph in Figure 9.13 agrees with the graph in Figure 9.12 insofar as they both imply that *Gender Ability Beliefs* screen off *Girls' Math Achievement* from *Teacher Math Anxiety*. However, if we assume the CTP, then the graph in Figure 9.13 is incoherent. Two different direct causal relationships – that between *Teacher Math Anxiety* and *Gender Ability Beliefs* and that between *Gender Ability Beliefs* and *Girls' Math Achievement* – are represented as occurring within a single time slice. On its face, the graph implies that the causal influence is transmitted instantaneously, which violates the CTP. Moreover, the implication that the influence of teachers on their students is instantaneous is implausible even if one is willing to tolerate instantaneous causation in some cases.

We suspect that Beilock et al. assumed that the value of *Teacher Math Anxiety* was unchanged over the course of the year, so that the value of *Teacher*

responsibility to the data), regardless of whether there are alternative time-free models that are admissible for their case. If Mayo-Wilson's strategy does work in this case, then we seem to have more reason to think that the Temporalization Criterion of Admissibility puts non-trivial constraints on causal modeling in the social sciences.

Math Anxiety at any time could be substituted for its value at any other time. Hence, the arrow from *Teacher Math Anxiety* to *Gender Ability Beliefs* in the model need not represent instantaneous causation. *Teacher Math Anxiety* might thus be indexed at an earlier time, such that it can influence *Gender Ability Beliefs* over a finite amount of time. Similarly, we suspect that *Gender Ability Beliefs* was assumed to have already taken on its end-of-year value at some earlier point in time, such that the causal relationship between *Gender Ability Beliefs* and *Girls' Math Achievement* might operate over a finite amount of time as well.

In order to represent the above assumptions explicitly, we can add intermediate time indices for each variable in Figure 9.13. The data give us no indication of the time scale at which these causal relationships might operate, so let's work with an arbitrary "lag time" of Δt between each time index. If we represent the beginning and end of the year with $t = 0$ and $t = T_f$, respectively, the causal graph looks like Figure 9.14.

At first glance, the temporalized version of the causal model depicted in Figure 9.14 seems far more plausible than that in Figure 9.13, since both *Teacher Math Anxiety* and *Gender Ability Beliefs* exert their influence gradually over the course of the year, rather than all at once at the end of the year. But although the graph in Figure 9.14 respects the CTP and looks plausible on its face, it does not entail the statistical relations Beilock et al. accept with respect to the measured variables. Recall that *Teacher Math Anxiety* (measured at the end of the year) was associated with *Girls' Math Achievement* (measured at the end of the year), but that association ceased to be significant if one conditioned on *Gender Ability Beliefs* (measured at the end of the year). For this reason,

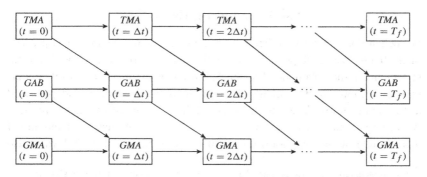

Figure 9.14 Temporalized version of Beilock et al.'s graphical model with intermediate time indices. (To consolidate the graph, we have shortened the variables names, where *TMA* stands for *Teacher Math Anxiety*, *GAB* stands for *Gender Ability Beliefs*, and *GMA* stands for *Girls' Math Achievement*.)

Beilock et al. drew a graph in which the influence of *Teacher Math Anxiety* on *Girls' Math Achievement* goes through *Gender Ability Beliefs*. By the same reasoning, GAB $(t = T_f)$ should screen off TMA $(t = T_f)$ from GMA $(t = T_f)$. That is, TMA $(t = T_f)$ and GMA $(t = T_f)$ ought to be independent conditional on GAB $(t = T_f)$. But that conditional independence relation is not entailed by the graph in Figure 9.14.[22] According to the graph in Figure 9.14, we would expect—consistent with Beilock et al.'s modeling practice – *TMA* at *any* time point (after the first two) to be associated with GMA at that same time point, conditional on GAB at that time point, because *TMA* and *GMA* are effects of a common cause from an earlier time-slice. For example, *TMA* at $t = 2\Delta t$ and *GMA* at $t = 2\Delta t$ are associated conditional on GAB at $t = 2\Delta t$, since *TMA* at $t = 0$ is a common cause of *TMA* at $t = 2\Delta t$ and *GMA* at $t = 2\Delta t$, and neither path is blocked by GAB at $t = 2\Delta t$.

One might defend Beilock et al. on the grounds that some parameterization of the model in Figure 9.14 might be able to represent the statistical facts. For example, if the time-indexed GAB variables are highly correlated, then GAB $(t = T_f)$ might screen off TMA $(t = T_f)$ from GMA $(t = T_f)$.[23] Hence, Beilock et al.'s model appears to be better than the other models we have considered. However, defending Beilock et al. in this way comes with a cost. Adjusting the parameterization in the way suggested makes the model unfaithful to the data, which appears to violate a modeling assumption that Beilock et al. themselves adopt.[24] One might worry, as well, that without some independent evidence for the stability of GAB over time, adjusting parameters in order to capture an independence not entailed by the graphical structure is ad hoc.

9.4 Concluding Remarks

We began this chapter by reflecting on the platitude that causes must precede their effects. We developed our Causal Temporality Principle (CTP) as a clear statement of that platitude. We then showed how time-free causal models may be temporalized in many different, incompatible ways, and proposed a Temporalization Criterion of Admissibility for time-free causal models. According to our criterion, a time-free causal model is admissible if and only if there is at least one temporalization of the model that satisfies CTP while also

---

[22] The problem remains if one assumes that the value of *TMA* remains constant for all time points. GAB $(t = T_f)$ does not screen off *TMA* at earlier time steps (e.g., $t = 0$) from *GMA* $(t = T_f)$. Similarly, the size of Δt is arbitrary and does not affect the result.

[23] Thanks to David Danks for pressing this point.

[24] Without the faithfulness assumption or something similar, Beilock et al. would have no reason to remove the edge from *TMA* into *GMA*.

being responsible to the data. And we discussed three cases from the social science literature that fail to be admissible according to our criterion.

Our overall goal has been to emphasize a disconnect between what researchers say about the value and importance of time for causal modeling and the way information about time is abused or neglected in practice. Our Temporalization Criterion of Admissibility is a minimal modeling standard, not a severe one. And yet, it is not difficult to find examples that violate the criterion in a variety of ways.

We hope that our work will at least have the salutary effects of convincing researchers in the social sciences to be more mindful of time when proposing time-free causal models and of alerting policymakers who use the work of social scientists to a potential source of error when interpreting such models. More ambitiously, we hope that social scientists will spend more time reflecting on the relationship between their theoretical commitments and their modeling practices. Causal modelers ought to explicitly articulate the assumptions that they make (or need to make) in practice and justify those assumptions. If the assumptions we have discussed in this chapter are not the right ones, then we hope that better replacements may be found.

In concluding our chapter, we want to remark on some remaining problems and directions for future research. Our Criterion of Admissibility is most naturally understood as a test of model adequacy. It would be good to transform or develop our Criterion into a constraint on methods of model building, rather than on models themselves.[25] Ideally, one would show that if a method satisfies some criterion, then all of the models the method delivers (or a sufficiently large proportion of them) will satisfy our Criterion of Admissibility. One might try supplementing existing automated causal search algorithms with our Causal Temporality Principle – as Glymour (2001) and Spirtes et al. (2000) suggest. But it is not obvious how to do so, since the CTP is not as epistemically transparent as principles that appeal to measurement times. Hence, finding a suitable temporalization criterion for methods appears to be an interesting and non-trivial direction for future research.

A related direction for future research is to determine the conditions under which various classes of causal modeling method produce correct models when applied to time-aggregated data and to find ways of testing for whether those conditions are satisfied in practical settings. Here there are serious problems having to do with the relationship between causal models for dynamic and static (equilibrium) systems, which have not yet been adequately explored by philosophers interested in causal modeling.

[25] Thanks to David Danks for raising this issue.

Acknowledgments

This chapter benefitted from helpful comments from many colleagues, including Jennifer Carr, David Danks, Samuel Fletcher, Konstantin Genin, Clark Glymour, Daniel Malinsky, Conor Mayo-Wilson, Shannon Nolen, John Norton, Federica Russo, Naftali Weinberger, audience members at talks given at the LSE, the University of Western Australia, and the TaCitS conference, and several anonymous referees.

10

Reintroducing Dynamics into Static Causal Models

Naftali Weinberger

Recently developed graphical causal modeling techniques significantly downplay the role of time in causal inference. Time plays no role in the criteria specifying what it means for causal hypotheses to be observationally equivalent, and the probabilistic criteria used fail to distinguish among hypotheses that – given the assumption that causal variables precede effect variables – involve different time orderings among the variables. Additionally, the causal Markov condition – a central condition for choosing among causal hypotheses given a joint probability distribution – most straightforwardly applies to cases in which the variables are sampled from time-stationary distributions. Finally, it is commonplace to present models in which the variables are not explicitly indexed to times.

The lack of emphasis on time in causal models may suggest that the models are neutral with respect to the temporal relationships among the causal variables. Here I propose that, in fact, the lack of reference to time in many causal models is a legacy of the fact that causal models were initially designed to model the relationships in simultaneous equations models. Simultaneous equations represent the stable long-term relationships among the modeled variables rather than the shorter-term dynamic of systems in which variables have not reached stable values. Rather than being neutral with respect to the temporal relations among the variables, many apparently atemporal causal models in fact apply only under specific temporal assumptions about the timescale at which the system is being modeled.

After spelling out the interpretation of causal models as applying to systems that are treated as static, I consider ways that these models can be generalized to model the dynamics of systems that are away from equilibrium. First, I show how the dynamic causal models developed by Iwasaki and Simon (1994) generalize the causal ordering method from Simon (1977), which had been designed for simultaneous equations. Simon's causal ordering serves as a basis for understanding the causal asymmetry of the structural equations that are still used to represent the dependence of effects on their causes. Second, I consider ways that the statistical methods used for causal modeling have primarily dealt with stationary distributions, and explain how certain non-stationary features of distributions correspond to the qualitative features in dynamic causal models. While I do not here provide a systematic treatment how to causally model time-series data, the present discussion contains both novel suggestions as to why such data present a challenge as well as suggestions for how dynamic causal models can be used as a bridge between contemporary graphical models and econometric time-series methods.

This chapter is organized as follows. Section 10.1 provides background regarding graphical causal models. Section 10.2 presents challenges to attributing times to the variables in these models, and explains how we can interpret the relationships among simultaneously measured variables causally. Section 10.3 discusses Iwasaki and Simon's dynamic causal models as a way of relaxing the presuppositions of simultaneous equations models. Section 10.4 draws connections between these dynamic causal models and methods employed in time-series analysis. Section 10.5 concludes.

10.1 Graphical Causal Models

The starting point of the present discussion is the graphical causal modeling methods developed by Pearl (2000) and Spirtes et al. (2000). While these works are relatively recent, their authors build on almost a century of contributions by psychologists, social scientists, economists, philosophers, and computer scientists. These methods provide a systematic and general account of how to use one's background causal assumptions to choose among competing causal hypotheses based on one's evidence, both in experimental and non-experimental contexts. For example, these methods enable one to determine which probability distributions over a set of variables are compatible with a particular causal hypothesis relating those variables. The methods also allow one to give a general account of when a causal quantity is *identifiable*, which means that it can be uniquely determined from the probability distribution, given one's causal assumptions.

In the graphical approach, causal models are typically represented using directed acyclic graphs, or DAGs. A DAG is a set of nodes connected by directed edges (i.e. arrows) such that one cannot get from a node back to itself via a set of connected edges all going in the same direction. When

DAGs are used for causal inference, the nodes are variables, the directed edges are direct causal relationships, and the lack of cycles corresponds to the assumption that no variable is either a direct or indirect cause of itself. The acyclicity requirement can sometimes be relaxed (Richardson, 1996; Park and Raskutti, 2016), although the general properties of cyclic graphs and the data-generating mechanisms they represent are less well understood. The term *causal model* typically refers to the combination of the graph and the joint probability distribution over the variables. As we will see, given a graph and distribution, it is sometimes possible to identify the functional relationship representing how an effect depends on its causes. As to what it means to interpret directed edges as direct causal relations, there are two general approaches. One is to treat the concept of direct cause as an undefined primitive whose content is determined holistically by the assumptions one uses to link DAGs to probability distributions. The other is to explicate X's being a direct cause of Y in terms of the possibility of changing Y via intervening on X while holding other causes of Y fixed.

I'll now provide further background regarding the use of DAGs for causal inference and identifiability, beginning with the latter. Let's begin by considering the simplest case in which a causal effect is *not* identifiable. Suppose that we were told that X causes Y, but these variables share an unknown common cause. The effect of X on Y would not be identifiable, since no matter how much we learned about the probabilistic relationship between X and Y, we could not know to what extent an inferred probabilistic dependency should be attributed to the causal relationship rather than to the common cause. In other words, one cannot uniquely determine the magnitude of the effect of X on Y from the DAG and probability distribution. In contrast, if we knew that X and Y shared a common cause Z, and that there were no other common causes of X and Y (other that causes of Z), then the effect of X on Y given particular values of Z could be identified by conditioning on different values of Z. Given the graph and the probability distribution, the effect would be identified by the expression $Pr(Y|X, Z)$.

A *path* is a set of connected arrows. Where X and Y share an unmeasured common cause, the reason the effect is not identified is that in addition to the single-arrow path from X to Y there is also an additional path between X and Y via the common cause that 'transmits' probabilistic influence. Given an account of the conditions under which probabilistic influence is transmitted, we could provide conditions under which the effect of X on Y can be identified. We have already seen that paths containing a common cause transmit probabilistic influence unless the common cause is conditioned upon. In contrast, causally independent causes of a common effect are uncorrelated *unless* one conditions on their common effect. So paths containing a common effect will not transmit probabilistic influence unless the effects are conditioned upon.

We can generalize these observations using the notion of *d-separation*:

Definition 10.1   d-Separation: A path is d-separated by variable set Z just in case:

(a) The path contains a triple $i \rightarrow m \rightarrow j$ or $i \rightarrow m \leftarrow j$ such that m is in Z, or
(b) the path contains a collider $i \rightarrow m \leftarrow j$ such that m is not in Z and no descendant of m is in Z. [m is a *descendant* of Z if there is a path from m to Z consisting of directed arrows all going in the same direction.] (Pearl, 2000)

d-separation is a property of paths. Two variables are d-separated by Z if and only if they are d-separated by Z along all paths. Two variables are d-connected by Z if and only if they are not d-separated. To see how d-separation helps with identifiability, note the following sufficient condition. The effect of X on Y is identifiable given variable set Z if, in the DAG, X is d-separated from Y by Z along any "back-door" path that contains an arrow going into X. While this is just a sufficient condition, there exists a sound and complete procedure for determining whether a causal quantity is identifiable from a graph and probability distribution.

Identifiability concerns what one can measure *given* causal knowledge and the probability distribution. But how do we get such knowledge in the first place? Here d-separation helps as well. We can determine which DAGs are compatible with a probability distribution using a criterion called the *causal Markov condition*, which states that d-separated variables will be probabilistically independent. More precisely:

Definition 10.2   Causal Markov Condition (CMC): given a graph G and a probability distribution P over V, for any sets of variables X, Y, and Z in V, if X and Y are d-separated conditional on Z in G, then X and Y are independent conditional on Z in P.

CMC will not hold generally if one considers a variable set V that omits common causes of variables in V. For ease of explication, in what follows I'll limit myself to considering *causally sufficient* variable sets that include any common cause of variables in the set. Causal sufficiency will not hold in general, and can be relaxed in many contexts. A further constraint is that the variables in V must be logically, conceptually, and metaphysically independent of each other. To see why, imagine two variables X and Y such that $X = A\&B$

and $Y = B$. Given this logical relationship, the probability of Y given X will be 1 even if the unconditional probability of Y is not 1. But this correlation reflects the logical relationship among the variables and does not by itself call for a causal explanation.

CMC specifies what it means for a DAG to be compatible with a probability distribution. The condition places only a weak constraint on the set of models compatible with a distribution and supplemental conditions are needed to choose amongst the hypotheses compatible with a given distribution. To see how weak the condition is, note that any complete graph in which all the variables are directly connected is compatible with any distribution (since no variables are d-separated conditional on any others). Nevertheless, to draw causal inferences from probabilistic knowledge *some* principle is required, and CMC's logical weakness gives it the virtue of being compatible with a wide range of accounts of causation. Moreover, given suitable restrictions on the variables considered (see above, as well as (Hitchcock, 2016, section 3.2)), it is a plausible candidate for a universally true generalization. The most serious potential counterexamples arise in quantum mechanics, and the interpretation of these cases remains controversial. I discuss some further alleged counterexamples in Section 10.4.

Using CMC combined with additional parsimony principles (Zhang, 2012; Forster et al., 2017) one can choose among causal DAGs based on a probability distribution. Yet there is a limit on how much one can infer from the distribution using these principles. Specifically, all DAGs that entail all and only the same conditional independencies according to CMC are considered to be observationally equivalent. For instance, one cannot distinguish between the three DAGs in Figure 10.1. As a result, additional background knowledge is required to distinguish among the DAGs, and a common plausible suggestion is that time will often play this role. Nevertheless, the fact that time is only invoked to choose among already specified models makes it clear that, in theory if not in practice, time plays a secondary role in graphical causal models. Notice, for instance, that if we were to know the time ordering of the variables X, Y, and Z in Figure 10.1, and to assume that causes preceded their effects, then the three DAGs would *not* be observationally equivalent.

The basis for using DAGs to establish identifiability is closely related to the metaphysical assumption underlying CMC. For example, blocking all back-door paths allows for identifiability because doing so blocks any way that the value of a cause is informationally relevant to the value of the effect variable, except via causal paths involving connected sets of unidirectional arrows from the cause to the effect. But this would not establish identifiability unless one believed that there must be some back-door path linking the cause to the effect

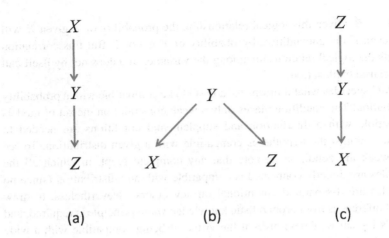

Figure 10.1 Three 'observationally equivalent' DAGs.

in order for one to provide information about the other that is not due to the causal path. In other words, one must accept CMC.

The concept of identifiability is crucial for understanding the relationship between causation and probabilities. The DAG framework does reduce causation to probabilities, but, given suitable causal assumptions, one can interpret certain conditional dependence statements as providing an unbiased measurement of the dependence of an effect on its causes. In such statements, the effect is the probabilistic consequent and its causes the antecedents in the probabilistic expression. When CMC holds, the joint probability distribution can be decomposed into a set of conditional independencies in which each variable is given as conditional only on its direct causes in the model. A useful way for thinking about the relationship between conditional dependence statements that can and cannot be interpreted causally is that only the former are invariant to interventions (Korb et al., 2004; Pearl, 2000). An intervention on a causal variable influences it in such away that any influence the intervention has on the intervened upon variable's effect is only via the intervened upon variable. The concept of an intervention allows for a systematic distinction between causal and non-causal probabilistic expressions, and has the added advantage of linking causal inference to experimental methodology, which also relies on interventions. Yet physically intervening is not necessary for gaining causal knowledge. What matters for whether a probabilistic quantity may be causally interpreted is whether it is identifiable, with or without intervening.[1]

One of the main purposes of causal models is to aid causal inference. CMC is evidently not sufficient for inference, since further parsimony principles are needed to choose among the large set of Markovian models. Additionally, most variable sets do not obey CMC (due to omitted common causes, for instance). When it comes to the best methods for inferring causal models from different types of data, a healthy dose of pragmatism is warranted. One can test an inference method for reliability by running a simulation and seeing if the method recovers the data generating process. Such a method can be reliable even if it does not always produce models obeying CMC. Nevertheless, the assumption that, given suitable restrictions on acceptable variable sets, the true DAG over such variables will either satisfy CMC or can be embedded in a broader variable set that satisfies CMC is indispensible for causal inference. The reason is the close connection between CMC and identifiability. We need identifiability to distinguish between probabilistic expressions that do and do not provide unbiased effect measurements. Without some such notion, there is no theoretical link between causal hypotheses and probabilistic information. Such a theoretical link is necessary even for running simulations. To talk about causal models as a data-generating process, one needs to specify the rules by which causal models generate the data. This is so even if one adds distortions to the data-generating process such that the inferred distribution will not itself obey CMC.

For CMC to play a theoretical role in providing identifiability conditions, it is a virtue that it provides only weak constraints on which models it deems acceptable. Note that the concept of identification considered here is nonparametric. That is, the claim that effects of X and Y on Z are identified from the probability distribution $\Pr(Z|X,Y)$ does not indicate which probabilistic expression or structural equation is appropriate for quantifying the effect of X and Y on Z. Whether, for example, the structural equation giving Z as a function of X and Y involves an interaction term (e.g. δXY) is not determined by the DAG, but the identifiability of the $Pr(Z|X,Y)$ ensures that in principle whether X and Y interact could be established from knowledge of the true probability distribution. The fact that the form of identifiability that CMC underlies is silent with respect to, e.g., which correlation coefficient is best for quantifying causal strength, provides a reason for thinking that the notion

---

[1] If we focus on 'hard' interventions that fully determine the value of the intervened upon variable, the effect of X on Y is identifiable by intervening on X if $\Pr(Y|X)$ is identifiable in the DAG derived from breaking all arrows going into X.

of probabilistic independence invoked by CMC should also be given in a way that is independent of any way of measuring probabilistic independence. As we will see in Section 10.4, there are alleged counterexamples to CMC with variables that are by hypothesis not causally related, but which appear to be probabilistically dependent if one uses certain correlation coefficients. There I will side with those who claim that these correlations are insufficient for establishing probabilistic dependence, and that whether two variables are genuinely dependent for the purpose of applying CMC is something that is to be determined by statisticians using whichever measure of dependence is appropriate for the data.

This brief discussion of graphical causal models is sufficient for seeing how they represent causal relationships without explicit reference to time. The asymmetric relationships in the models are not explicated in terms of causes preceding their effects in time, but rather in terms of the different probabilistic consequences of the directions of the arrows in a DAG. In Section 10.3 we will see that the project of explicating causal directionality in fact precedes probabilistic approaches, and in Section 10.4 we will consider some subtle ways in which probabilistic assumptions relate to temporal ones. But before we get there, we need to ask an even more basic question. How do we make sense of causal models that do not specify the temporal relations among their variables?

10.2 Temporal Relations in Seemingly Atemporal Causal Models

In the previous section we reviewed graphical causal models and saw how the basic causal concepts are defined without explicit reference to time. Yet on reflection, it is somewhat puzzling as to how to understand the variables within the causal models without reference to time. In many causal models there is a single variable for each quantity. For instance, a model for an ideal gas system might have a variable V for volume, and not, for instance, a vector of variables $V_1, V_2, \ldots, V_n$ for the volumes at times 1 to n. In some cases, this does not lead to ambiguity. In the ideal gas case with a moveable piston, we know that equilibrium volume depends on pressure and temperature, and so we do not need to consider values away from equilibrium. But, more generally, simply to talk about the relationship between X and Y without saying when Y occurs (at least relative to X) leaves it ambiguous which relationship we are talking about. So we need to ask: which assumptions make it appropriate to compress potentially complex sets of causal relationships among time-series variables into simple representations involving single variables?

Some philosophers may take issue with my claim that the variables in a causal model in fact have a temporal ordering. Philosophers commonly treat variables as referring to properties, where the relationship between a variable and its values is akin to that between a determinable and its determinates (e.g. the relationship between 'being a shade of red' and 'being crimson'). On this way of talking, a variable can have different values at different times, since an object's properties can change over time. Yet this way of talking does not match statisticians' definition of a variable. Mathematically, a variable is a mutually exclusive and jointly exhaustive partition over a space of possibilities. Since it follows that a variable can have at most one value, to represent a quantity that changes over time we need multiple variables – typically, one corresponding to each observation. If we are not talking about a quantity that is changing in time, then we may define a variable without respect to time. But since we are here considering the role of time in causal models, it will help to eliminate ambiguity and to explicitly assign each variable a time.[2] Where we want to talk about a quantity changing over time, we will need a time-series of variables: $\{X_t, X_{t+1}, \ldots, X_{t+n}\}$.

It is natural to assume that once we make the time-indices of the variables in a causal model explicit, the temporal ordering should match the causal ordering. That is, causes must precede their effects. In the present volume, Livengood and Zwier's Chapter 9 seek to reconcile this plausible assumption with actual modeling practice. Here I propose an alternative explanation for why graphical causal models often do not include explicitly time-indexed variables. Specifically, all of the variables are indexed to the *same* time. This, I suggest, is a legacy of origin of certain aspects of the causal modeling framework in an econometrics tradition that uses so-called simultaneous equations. I will say a bit more about that tradition in the next section. Here I will simply assume that the variables in many causal models are simultaneous, and consider how we might interpret the relationships among such variables causally.

Talk of causal relations among simultaneous variables raises a puzzle, as causes are generally assumed to precede their effects. Some philosophers do allow for simultaneous causation, and equilibrium relationships – such as the effect of the current supply and demand of a good on its current (equilibrium) price – are plausible candidates for simultaneous relationships. Yet there are available causal interpretations of simultaneous equations, even assuming that causal relationships are genuinely diachronic. I consider two: the 'limit'

---

[2] See (Hitchcock, 2012) for a further defense of time-indexing variables.

interpretation and the 'steady-state' interpretation.[3] On the limit interpretation, the claim that X causes simultaneous variable Y should be interpreted as saying that X influences Y diachronically over arbitrarily small intervals. On the steady-state interpretation, which I adopt, simultaneous causal relationships among variables should not be understood as genuinely simultaneous, but rather as simultaneously observed. To interpret the relationship between X_t and Y_t causally one needs to assume that (a) X_t equals X_{t-1} and (b) Y is in a steady state at t. Condition (a) indicates that X_t itself is not the cause of Y_t but nevertheless enables one to infer the value of the genuine prior cause. Condition (b) indicates that Y at t has had sufficient time to reach a steady state in response to any changes in X. A variable is in steady state just in case its time-derivative is 0.

The steady-state interpretation of simultaneous equations is somewhat counterintuitive. While talk of simultaneity suggests that one is evaluating a system at a very short timescale, on the suggested interpretation simultaneous equations represent long-run relationships. To avoid confusion, it is important to distinguish between two inversely related rates that we might be referring to when talking about time in causal models. On the one hand, we might be talking about the timescale at which we evaluate a system, operationalized in terms of the rate at which the system is sampled. Here slower rates of sampling correspond to evaluating the system at a longer timescale. On the other hand, there is the rate at which a variable stabilizes in response to changes in its causes' values. One can think of simultaneous equations as evaluating the system at a sufficiently long timescale such that the rate at which the effects in the model respond to their causes is negligibly small at the sampling rate corresponding to that timescale.

Why interpret non-time-indexed causal models in the manner suggested here? Suppose that X causes Y, but it takes time for Y to stabilize. To make this a bit more concrete, imagine that if $Y = y_0$ at t and X is set to x at t, then Y will equal y_1 after 1 second and y_2 after 2 seconds, and will then remain at y_2 in the absence of further interventions. If we modeled this as $X \rightarrow Y$ it would be unclear at which time Y was being measured and we could not give a quantitative structural equation. There are two options: either X and Y are not represented as simultaneous or they are. If not, then the modeler is being sloppy, and the sloppiness can only be eliminated by specifying how much time has passed between X and Y. If X and Y are simultaneous, then we need to choose between the limit and the steady-state interpretation. On the limit interpretation the effect of X on Y corresponds to the instantaneous rate of

---

[3] This discussion is inspired by (Strotz and Wold, 1960, section 3).

change of Y given X at t: $\frac{dy}{dx}$. This fails to capture the longer-term effect of X on Y, and especially that Y is constant after t=2. In contrast, the steady-state interpretation clarifies that the variables are measured simultaneously, but does not require one to know precisely how long it takes Y to fully respond to the intervention on X. It indicates simply that the system is being evaluated at a timescale at which Y has had enough time to respond. So assuming that causal relationships take time and the modeler is not sloppy, treating the variables as simultaneous is the natural way to go.

The proposal that the variables in non-time-indexed DAGs are simultaneous illuminates some otherwise puzzling features of contemporary causal modeling. It would explain, for instance, why so much of causal inference is done on cross-sectional rather than longitudinal data, and also why social scientists working with time-series data often reach for alternative causal frameworks such as those relying on Granger causality (though see (Eichler, 2012) for a discussion of the use of DAGs for time-series analysis). This suggests that existing models have limitations regarding how to model time, though it is surprisingly unclear what those limitations are. The strategy of the present chapter is to begin with the working hypothesis that the limitation of causal models is due to their original application to simultaneous equations with the steady-state interpretation, and then to consider how they can be generalized to systems that are away from steady state. The following section presents Iwasaki and Simon's dynamic causal models, which provide such a generalization. These are, to be clear, only one type of dynamic causal model, and I here make no attempt to provide a survey of the options. Nevertheless, the following discussion will reveal them to be especially useful for thinking about the temporal assumptions of causal models.

10.3 Reintroducing Dynamics into Causal Ordering

The dynamic causal models in Iwasaki and Simon (1994) are a generalization of Simon's causal ordering method (Simon, 1977). While Simon does not make this explicit in the earlier paper, Iwasaki and Simon later write:

> [T]he theory's definition of causal relations is useful in generating a causal interpretation of the behavior of any system described by a system of simultaneous equations, including a physical device. Causal ordering was initially defined by Simon for a static system consisting of equilibrium equations. (1994, p. 145)

The progression in this section is from causal models with the steady-state interpretation to models that can represent systems away from steady state. I will begin by presenting Simon's original causal ordering method and explaining its relationship to contemporary modeling methods. The fact that his account can still be used to explain the asymmetry of the structural equations in causal models supports the working hypothesis that seemingly atemporal causal models contain simultaneous variables.

Simon asks what makes causation an asymmetric relationship, and he rejects the answer that it is temporal ordering. His strategy is to show how one can begin with a set of 'symmetric' equations in which one can freely move terms across the equals sign via applying the same operation to both sides, and to provide conditions under which such equations can be reorganized so that each variable is given as an effect of its causes. Systems that can be interpreted causally are those in which the values of certain variables can be solved for prior to others, and the causal ordering – that is, the partial ordering of the variables such that causes always precede their effects – is determined by the order in which the variables must be solved for.

As a simple example, consider an ideal gas system with the variables temperature (T), pressure (P), and volume (V). If the relevant system is a sealed container, we begin with the following symmetrical equations:

$$0 = f(P, V, T) \tag{10.1}$$

$$0 = f(T) \tag{10.2}$$

$$0 = f(V) \tag{10.3}$$

Equation (10.1) indicates that there is a function relating temperature, pressure, and volume, i.e. the ideal gas law. Equations (10.2) and (10.3) indicate that the values of temperature and volume are *exogenous* – that is, they do not depend on those of other variables in the system. Given these equations, one can solve for T and V first, and only then can one solve for P in equation (10.1). Accordingly, the equations can be rewritten as follows:

$$P = f(V, T) \tag{i}$$

$$T = f(t) \tag{ii}$$

$$V = f(v) \tag{iii}$$

Equation (i) indicates that pressure is an effect of volume and temperature. Equations (ii) and (iii) indicate that T and V are assigned constant values. After reorganizing the equations, one can treat them as structural equations in which the variables on the left-hand side are effects of those on the right-hand side, where each variable appears on the left-hand side in a single equation.

Of course, not every set of symmetric equations can be causally ordered in this way. If one could not divide up the equations so that certain self-contained subsets could be solved first, no causal ordering could be identified given the equations.

In one sense, the causal ordering method does not provide causal information beyond that required for specifying the equations in the first place. We needed to specify which variables are lawfully related (in (10.1)) and also which variables were exogenous to find the causal ordering. If pressure, instead of volume, were exogenous, we would have to replace (10.3) with an equation (10.4):

$$0 = f(P) \tag{10.4}$$

These equations entail that T and P are causes of V. This causal ordering represents the relationships in a movable piston system, rather than a fixed-volume container. Like recent methods, the causal ordering method does not yield causal knowledge without causal assumptions. Nevertheless, the method reveals how causal asymmetries follow from symmetric laws and assumptions about exogeneity.

Although Simon's method does not rely on probabilities, the structural equations identified using the methods are the same as those identified using graphical causal models. To see how this could be, it helps to know that any variable set linked by equations with a deterministic component plus an independent error term for each variable automatically satisfies CMC. Accordingly, exogenous variables can be operationalized as those whose values depend on an independent error term alone. Nevertheless, we may be able to make judgments of independence even in cases where we cannot make probability attributions. In the ideal gas case, the judgment that temperature is exogenous is due to the ability of an experimenter to change the temperature at will by adjusting the heat bath in which the container is immersed, and the judgment that volume is exogenous reflects the assumption that the volume is either fixed by the experimental setup or can be adjusted by the experimenter. While modeling external sources of variation in a system as random variables enables one to infer causal relationships from probabilistic information, the type of independence that underlies causal attributions need not be identified with probabilistic independence.

I now consider Iwasaki and Simon's dynamic causal models, which generalize Simon's causal ordering method to systems in which at least some of the variables have not yet reached steady state. I will especially emphasize the models' use of time-derivatives to expand the representational power of the framework. This will help clarify the models' temporal assumptions, and will

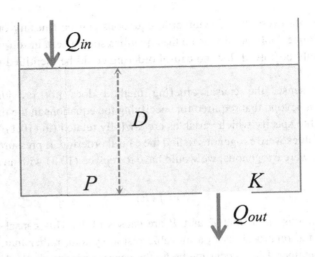

Figure 10.2 Iwasaki and Simon's bathtub example: Q_{in}: *rate of flow in*, Q_{out}: *rate of flow out*, D: *depth*, P: *pressure*, K: *drain size*.

further allow us to later compare Iwasaki and Simon's use of derivatives to the somewhat similar use of derivatives in time-series econometrics models.

Let's begin with an example (Figure 10.2). Consider a bathtub in which water flows in at a rate of Q_{in} and out at a rate of Q_{out}. The difference between these rates determines the rate of change of depth, which in turn influences depth, D. The pressure P at the bottom of the tub is proportional to depth. The rate of flow out depends on pressure and the size of the drain, K, which is treated as exogenous. The system is at steady state when Q_{in} equals Q_{out} and D no longer changes.

The symmetric equations for the steady state of this system are as follows.

$$Q_{out} = c_5 k P \tag{10.5}$$
$$D = c_6 P \tag{10.6}$$
$$Q_{in} = c_7 \tag{10.7}$$
$$K = c_8 \tag{10.8}$$
$$Q_{in} = Q_{out} \tag{10.9}$$

Equations (10.5) and (10.6) indicate that the rate of flow out is proportional to the size of the drain and pressure, and that depth is proportional to pressure. Equations (10.7) and (10.8) say that the rate of flow in and the size of the drain are exogenous. Equation (10.9) indicates that at equilibrium the rate of flow in equals the rate of flow out. This is what indicates that we are considering the

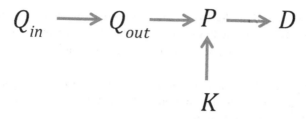

Figure 10.3 Equilibrium causal ordering for bathtub.

model at equilibrium. The resulting causal ordering is given by the graph in Figure 10.3. Since Q_{in} is exogenous, one can solve for Q_{out} via (10.9). From (10.5) we see that given K and Q_{out} we can solve for P, and then from (10.6) we derive D.

The causal model in Figure 10.3 is far from intuitive. To make sense of it, it is crucial to bear in mind that we are assuming that all variables have reached stable values in response to any perturbations of their causes. While changing the rate of flow in would have a short-term effect on depth and pressure, assuming that the system reaches steady state, the long-term rate of flow out will only depend on the new rate of flow in. Similarly, changing the size of the drain would alter the steady-state pressure and depth, but would not influence long-term Q_{in} and Q_{out}. This is not to say that the model is wholly satisfying. It is especially difficult to understand the causal ordering of pressure and depth. Within the causal ordering method, the way to understand this is that assuming that the system is at steady state, and that the drain is size k, the steady-state pressure must have a certain value p in order to achieve the required rate of flow out, and in order for pressure to have this value the water must be a certain depth. But this seems to describe teleological rather than causal relationships. In fact, Iwasaki and Simon bring this example in large measure to motivate the desirability of more dynamic representations of a system, which yield a more intuitive ordering. Whether or not the model is intuitive, it bears mentioning that Dash (2003) uses simulations to confirm that it is the model we would infer when sampling at a longer timescale.

Since Q_{in} and Q_{out} are rates of change, we could have represented them using time-derivatives. But we need not do so. While the traditional causal ordering method applies to variables that have reached stable values, these variables can themselves be derivatives, in which case the stable values are constant rates of change. In extending the framework to represent the dynamics, the innovation is not the use of time-derivatives per se, but the use of both variables and their time-derivatives within a single model. Since variables and their time-derivatives are conceptually related, they are not fully independent. But, as we will now see, including both variables and their time-derivatives in a model enables us to reintroduce dynamics into a static set of equations.

Naturally enough, if we no longer assume that the system is at steady state, we need to eliminate the assumption that Q_{in} equals Q_{out}. We can replace equation (10.9) with the following:

$$D' = Q_{in} - Q_{out} \tag{10.10}$$

Equation (10.10) states that the rate of change of depth is equal to the difference between the rate of flow in and rate of flow out. When adding an equation with an explicit derivative into our model, we also need to add two other equations. One states that integrating D' yields D:

$$D = Int(D') \tag{10.11}$$

The other specifies an 'initial condition' for D at time 0:

$$D_0 = d \tag{10.12}$$

These equations enable us to represent the passing of time in the model. Given a variable's current value and its time-derivative, integration yields a discrete approximation of the variable's value at subsequent moments. Equations such as (10.12) are crucial for the causal ordering method, since they indicate that at a particular time-step we need to treat the value of the non-equilibrated as exogenously given. Given such a value, the model will tell us how the variables will evolve over time.

The dynamic model derived from combining (10.5) – (10.8) with (10.10) – (10.12) is given in Figure 10.4. The water flowing in increases depth, which increases pressure, which, given a drain of size k influences the rate at which the water flows out. The rates at which the water flows in and out in turn determine the rate at which the depth changes and thereby influence the future depth.

The graph in Figure 10.4 provides a subtle way of integrating activities occurring at several timescales into a single model. Variables linked by causal arrows are interpreted as they were in simultaneous equations models. That is, P, K, and Q_{out} respond quickly to changes in their causes and are measured at steady state. If desired, one could include additional derivatives for all variables in the model, but this would not substantially change the causal relationships

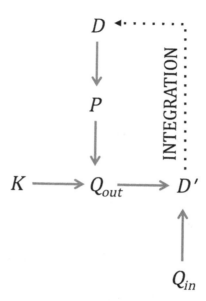

Figure 10.4 Dynamic causal ordering for bathtub.

among the variables.[4] However, the inclusion of D' along with the integration link indicates that we are sampling the system at a rate at which depth has not reached a stable value. Using this derivative, we represent how current depth influences its rate of change. It is via this form of self-regulation that the system gets moved towards equilibrium in the long run.

To get a feel for the dynamic graph, it helps to consider how we might represent the same causal relationships using a 'rolled-out' graph with time-indexed variables and no derivatives (Figure 10.5). While the rolled-out graph does not explicitly represent the way that the rate of change of D is a function of Q_{in} and Q_{out}, and that D' combines with D_0 to produce D_1, it otherwise preserves the causal ordering and differentiates between the synchronic and diachronic relationships. It helps to think of this graph in light of the CMC. Note, for example, that given D_1, P_1 is probabilistically independent of P_0. This reflects the sense in which P is 'memoryless' – to know its value at a time, we need not know its prior values. More generally, causal models for

---

[4] To illustrate, if we included P' in the model and added the required equations, the resulting graph would contain an arrow from D to P', an integration link from P' to P, and causal arrows from P to P' and from P to Q_{out}. This would indicate that it takes time for depth to influence pressure, but would not further change the relationships between D, P, and the other variables in the model.

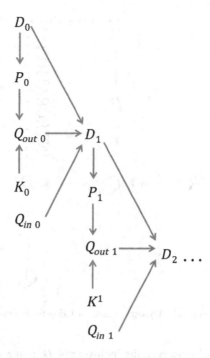

Figure 10.5 'Rolled-out' graph without time-derivatives.

static relationships are memoryless – which is why we needed to specify initial conditions only once we moved to the dynamic model. In contrast to pressure, depth is *not* memoryless when considering the system away from steady state. To know what the depth will be at the next moment you need to know its current value. As long as the system is away from equilibrium, the value of depth serves as a record of a past event – often the event that shocked the system out of equilibrium – and as an index for how far the system is from equilibrium.

The relationship between Figures 10.4 and 10.5 resembles that between causal models for simultaneous equations and time-indexed causal models more generally. The causal ordering method yields shorthand representations of the temporal development of a system without presupposing a characteristic timescale. In the static case, the graphs indicate that the variables are at steady state without needing to specify precisely how long it took them to stabilize. Similarly, causal models with derivatives convey information both about the relative rates at which variables influence one another and about the timescale at which the system is being evaluated, without providing specific rates. From the inclusion of a variable along with its derivative, we can infer that the variable does not reach a stable state at the timescale at which the system is considered, and if there are other variables in the model without their derivatives, we know that they reach steady state at a comparatively faster rate.

Earlier I discussed the counterintuitive device of using simultaneous equations to model long-run relationships. The explanation I gave there similarly helps clarify how including diachronic causal relationships corresponds to evaluating the system at a shorter timescale. Specifically, by 'zooming in' and sampling the system at a faster rate, the slower relationships that appeared as simultaneous at the longer timescale now can be seen unfolding in time at the shorter timescale characterized by the faster sampling rate. While this is a lot to keep track of, the feature of the representation that is not counterintuitive is that synchronic relationships refer to faster causal interactions than those represented by diachronic relationships.

Iwasaki and Simon develop an 'equilibration' operator, which transforms a dynamic model into a model with the causal ordering one would get were one to wait for the equilibrated variable to reach steady state. To equilibrate D, delete equations (10.11) and (10.12) indicating the integration equation and the initial value of D, and replace D' with 0 in equation (10.10) ($D' = Q_{in} - Q_{out}$). Then use the causal ordering method to derive the equilibrium mapping. Since replacing D' with 0 in (10.10) yields that $Q_{in} = Q_{out}$, we end up with the same equilibrium equations we started with and therefore get the equilibrium causal ordering. More generally, to equilibrate X (Dash, 2003):

1 Set all derivatives of X in the model to 0 and remove them from the model.
2 Delete all equations going into X or its derivatives.
3 Remap to get the new causal ordering.

The second step deletes both the integration equation and that stating X's initial conditions. Note that equilibration is always relative to a variable or set of variables. In models with multiple dynamic variables, certain variables sometimes must be equilibrated before others (Iwasaki and Simon, 1994, p. 166). Note also that equilibration does not always preserve the causal ordering among the variables, as can be seen from the reversal of the arrows linking D, P, and Q_{out} between the equilibrium and dynamic models.

The link between Iwasaki and Simon's dynamic causal models and the econometric time-series we will consider in the following section is that both use simultaneous equations to represent long-run equilibrium relationships and diachronic equations to represent non-equilibrium behavior, and both rely on derivatives (or difference equations) to do so. The econometric tradition is, in general, much less systematic in identifying the causal assumptions of their models than the causal modeling tradition, but also more focused on issues of measurement, especially when it comes to the analysis of time-series data. So, the similarities just mentioned bode well for a better unification of causal theory with measurement methods.

10.4 Dynamic Causal Models and Econometric Time-Series Methods

While I suspect many philosophers and social scientists will be surprised by my claim that non-time-indexed models should be taken as representing steady-state relationships, there is one context in which philosophers have noticed that causal models are easier to apply in static scenarios. Specifically, for non-stationary time-series, more sophisticated statistical tools are needed to determine whether variables are probabilistically dependent, and thus, by CMC, d-connected in the true causal graph. This issue arises in discussions of Elliott Sober's proposed counterexample to CMC (Sober, 2001). The example involves the relationship between the sea level in Venice and British bread prices. These quantities are not causally related, but since both increase over time, the value of one is informative about that of the other. So, Sober claims, we have a case of a correlation without a causal explanation, violating CMC. I have deferred discussion of the counterexample until now, since it introduces time-series concepts that will help us understand the econometric methods that are the subject of the present section.

In my view, the correct response to Sober is that bread prices and sea levels are not in fact probabilistically dependent, and thus there is no dependency lacking a causal explanation. Hoover (2003) presents a helpful discussion of why the counterexample fails. In defining CMC in Section 10.1, we presumed knowledge of the true probability distribution over the modeled variables, and in particular knowledge of which variables are independent. But statistical independence is not directly observed; it must be inferred from a finite sample. Statistical inference requires a model of the distribution from which the data is sampled. One common assumption is that the random variables in a sequence are independent and identically distributed. If one uses the wrong probability model for statistical inference, one will make incorrect inferences regarding which variables are probabilistically independent, and thus regarding which causal models are ruled out by CMC.

In determining which probability model is apt for a set of observed random variables, it is crucial to consider whether the underlying distribution is stationary or non-stationary with respect to time. Informally, a sequence is non-stationary if its variables or the relationships between them exhibit a systematic dependence on time. Sober's sequences are manifestly non-stationary, since the mean values of the variables in the sequences increase with time. If one models the two time-series for bread prices and sea levels using a correlation coefficient designed for stationary time-series, one will get a positive value, but this will not reflect a genuine probabilistic dependency. As Hoover notes, econometricians would model the time-series in Sober's example as being $I(1)$, indicating that although each time-series is non-stationary, the time-series derived from taking the first difference between time-steps – that is, subtracting each variable from its predecessor – is itself stationary. (Taking the difference in this manner corrects for autocorrelation.) The process of taking the difference between time-steps is called *integration*, the n number of times one must take the difference to achieve stationarity is called the *order of integration* $I(n)$. For two $I(1)$ series such as that in Sober's example, the test of whether they are probabilistically dependent is whether there is a function of the two differenced time-series that is itself stationary. This roughly corresponds to whether one time-series is informative about the other, even once one has taken the difference. In Sober's case, the answer is presumably no. Although the *level* of the sea is informative about the level of bread prices, the *changes* in one are not informative regarding the changes in the other.

While I view Hoover's response to Sober as successful, it is only a starting point for a broader discussion of causal inference from time-series. Defenders of Sober's counterexample (Reiss, 2007) have argued that Hoover's response severely limits the usefulness of CMC in inference by limiting the variable sets to which the condition applies and making causal inference overly dependent on the prior work of statisticians who process the data. Yet, as I argued above, while CMC can be relaxed and modified to deal with variable sets not satisfying CMC, its role in providing identification criteria specifying when a probabilistic relationship is causally interpretable is indispensible. For this purpose, what matters is that variables cannot be mutually informationally relevant in the absence of some causal connection (i.e. some d-connected path) and it is ok for the causal modeler to outsource the task of establishing informational relevance to the statistician. That said, CMC is only part of causal inference and we should not infer from Hoover's discussion that, as a general strategy, causal modelers can relegate the nuances of time-series inference to the statistician. As we have seen, causal models incorporate very specific temporal assumptions, and these assumptions are closely related to those studied by time-series econometricians.

Let's think further about stationary and non-stationary time-series. A simple example of a time-series gives each variable in the series as a function of its prior value plus an error term:

$$x_t = \rho x_{t-1} + \epsilon_t \qquad (10.13)$$

When $|\rho| < 1$, the time-series given in (10.13) is stationary, and when $|\rho| = 1$ the time-series is non-stationary. This derives from the way that the shocks represented by the error terms accumulate. When $|\rho| < 1$, shocks from the past will continue to have an influence into the future, but the longer ago a shock is, the less influence it will have now. This is because although a shock now is incorporated into the present value of X, at each subsequent time-step the value of X gets multiplied by ρ. So the contribution of a shock now to the trend n time-steps later will be a function of ρ^n, which tends towards 0 as n increases as long as $|\rho| < 1$. In contrast, when $|\rho| = 1$ the force of each shock is never diminished. In stationary series, one can abstract away from the role of time in the sense that the relationship between the variables will only depend on their relative times, not absolute time. More specifically, in a stationary time-series, the mean and variance (and higher-order moments) are independent of time, and the covariation of variables at different times depends only on how far apart they are.

We see that even with stationary time-series a variable's current value reflects prior shocks, though as time passes the influence of prior shocks diminishes. Accordingly, ρ can be understood as indicating how much 'memory' a variable has of earlier events in the series. As we saw in our discussion of Iwasaki and Simon, considerations of how much memory a system has are not incidental to the content of causal models. While a system that has reached equilibrium retains no memory of the prior shocks that perturbed it from equilibrium, and can be represented using simultaneous equations, when a system is away from equilibrium a variable's value does contain information regarding prior shocks, and this must be incorporated into the model via relationships across time-steps. Whether one treats a stationary time-series as having memory depends both on the timescale at which one considers the system (i.e. the length of the lags) and a judgment regarding how small the influence of a prior shock needs to be before it counts as negligible enough to be ignored.

These considerations suggest that extending causal models from the simultaneous case to the dynamic one will require the incorporation of more sophisticated time-series methods into causal modeling. Even if causal modelers continue outsourcing the data analysis to statisticians, they cannot ignore that models for different timescales correspond to different time-series with distinct statistical properties.

Of course, the very simple time-series represented by (10.13) does not involve causal relationships. Let's now turn to a causal example. The following example comes from Hoover's illuminating discussion (Hoover, 2015) of when we might interpret the relationships in econometric time-series models causally. The example concerns a drunk who, after leaving the bar, is followed by a concerned friend. The friend following the drunk keeps a safe distance in order to avoid detection, but will speed up to get closer to the drunk whenever the distance between them gets too large.

The drunk's trajectory is a random walk. That is, at each time-step the drunk is equally likely to move in any direction. The equation giving the movement of the drunk is that in (10.13) with $\rho = 1$, meaning that the time-series for the drunk's trajectory will be non-stationary. The friend's movement is also represented by a non-stationary time-series. Nevertheless, the time-series for the friend and the drunk are first-order cointegrated. This cointegration relationship indicates that in the long term, the friend and the drunk will never drift more than a certain amount apart. Since the friend is following the drunk, it is clear that the position of the drunk causes that of the friend. But this relationship will only be evident when considering the system at the right timescale. If one observes the system at too short a timescale, one will miss the fact that the friend and the drunk never drift too far apart. If one observes it at too long a timescale, one will discover the long-term cointegration relationship, but miss that it is the drunk influencing the friend and not vice versa.

The causal ordering framework does a nice job representing this case (Figure 10.6). X_d is the position of the drunk and X_f is that of the friend. In the dynamic graph on the left, the distance between the drunk and the

(a)   (b)

Figure 10.6 Dynamic graph for cointegration example X_d: *position of drunk*, X_f: *position of friend*.

friend determines the position of the friend at the next time-step. The use of
a derivative/integration link pair indicates that the friend reduces the distance
only after some time has passed. (Note that the *cointegration* relationship
between the time-series is represented using an integration link in the causal
model.) Applying the equilibration operation to X_f yields the simpler graph
on the right, which indicates that the position of the drunk is causing that of
the friend.

The present example could be represented in econometrics using an error
correction model (Engle and Granger, 1987; Wooldridge, 2015). In the background, we assume that the time-series for the friend and drunk are I(1), and
that there is a long-run relationship between them as follows:

$$x_{f(t)} = \alpha + \beta x_{d(t)} \tag{10.14}$$

We represent the equation for the dependence of the position of the friend
on that of the drunk as follows:

$$\Delta x_{f(t)} = \delta \Delta x_{d(t)} - \gamma (x_{f(t-1)} - \alpha - \beta x_{d(t-1)}) + u_t \tag{10.15}$$

Equation (10.15) takes the first difference for both the friend and the drunk
to render the non-stationary time-series stationary. δ indicates the short-term
effect of the change in the position of the drunk on the change of position of the
friend. In our example δ will be 0, as the friend does not respond immediately
to the drunk's movements. The term in parentheses represents the longer-term
cointegrating relationship. Given the long-run equilibrium relationship from
(10.14), any difference between the first term and the difference between the
second and third term indicates that the system is away from equilibrium. The
coefficient γ determines how quickly the system will return to equilibrium.

It is straightforward to see how Equation (10.15) with $\delta = 0$ neatly
corresponds to the dynamic graph in Figure 10.6(a), and how equation (10.14)
corresponds to the static model in Figure 10.6(b). The example chosen was
exceedingly simple, so it remains to be seen how generally causal models
can be linked to econometric time-series models. Nevertheless, by linking
the dynamic causal model to a well-understood econometric model, we see
how the mathematical operations that econometricians use to correct for
autocorrelation map on to the relationships in dynamic causal models.

I conclude this section with a reflection of the relationship between causal
and statistical inference when it comes to more complex time-series. I've
argued that CMC is a general rule of causal inference and that the question
of which variables are probabilistically independent should be left to the
statisticians. Relatedly, if variables are probabilistically dependent, CMC does
not differentiate among the inferences one can draw from different types of

dependencies. But this just means that we need principles in addition to CMC. This is generally acknowledged even in the stationary case, where CMC must be supplemented by additional principles to choose among Markovian models. When we consider more complex time-series, we need to make additional distinctions among ways that variables can be probabilistically dependent, and whether we treat certain vectors of variables as dependent or independent will sometimes depend on the timescale at which we consider the system. This is the lesson of Iwasaki and Simon's dynamic causal models, and the present section provides preliminary confirmation that inferring these models from time-series data will require an appeal to time-series techniques from econometrics.

10.5 Conclusion

While graphical causal models have revolutionized causal inference and analysis, it remains surprisingly difficult to specify their domain of application. There is work to be done in determining how graphical methods can be used for inferring causal hypotheses based on time-series data. Yet in order to figure out how the framework can be expanded, we need to first get clear on its current limitations. Here I have suggested that far from being neutral regarding the temporal relations among causal variables, existing models often implicitly come with strong assumptions about the variables' temporal stability. This diagnosis indicates that generalizing the framework is a matter of expanding it to represent cases in which at least some variables are away from their steady states. I have shown how Iwasaki and Simon's dynamic causal models provide such a generalization, and further illustrated the relationship between the generalized representation and concepts from econometric time-series methods.

11

Overcoming the Poverty of Mechanism in Causal Models

David Jensen

This chapter, written some months after my talk at TaCitS, is my attempt at a "rational reconstruction" of the presentation I gave at the conference. It hews fairly closely to the organization of that talk. It may be slightly better, due to a few additional thoughts I've had in the intervening months (some stemming from the questions and discussion after the talk). It also may be slightly worse, due to the absence of things I said off-the-cuff during the event, inspired by other talks and coffee-break discussions.

I aimed for the type of invited talk that I particularly value, one that poses more questions than answers and that draws together ideas from multiple areas of scholarship. Instead of attempting to provide a magisterial overview of some body of work, I instead tried to imagine where a new body of work might go. In doing that, I hope to introduce readers to some new ideas that could influence their thinking in the future.

I want to thank Samantha Kleinberg and the other organizers for giving me the opportunity to address this community. The conference was extraordinary – an intellectual feast from start to finish. It was the sort of gathering you hope for when you join a scholarly community, but all too rarely find. It drew researchers from diverse communities, but all with highly complementary interests, goals, and expertise. Attending and speaking was a great pleasure, and it gave me the opportunity to crystalize a set of ideas that have been bubbling inside my brain for some time.

Both then and now, I am keenly aware that some of the topics here are substantially outside of my comfort zone and my areas of expertise. I come to this topic as a computer scientist, and (at best) an amateur philosopher and social scientist. I apologize in advance for any naiveté regarding the views, methods, and findings of researchers in those latter fields.

11.1 An Example

Before I describe the key ideas, let me start with an example that spurred some of my own thinking on this topic. In April 2017, Jayash Paudel, a graduate student in the Department of Resource Economics at my university, presented a talk to my research group. Jayash, together with his advisor Christine Crago, had recently completed an analysis of government subsidies and agricultural productivity in Nepal. Specifically, they studied the impact of a government program that subsidized fertilizer purchases in a specific region of Nepal and that aimed to enhance agricultural yield for farmers engaged in subsistence agriculture (Paudel and Crago, 2017).

The conclusions of the study were surprising: "Using data from household surveys conducted before and after the program, we exploit eligibility criterion to show that the subsidy, on average, leads to a significant decrease in the use of chemical fertilizers and annual agricultural yields." That is, subsidizing the use of chemical fertilizers *decreased* both fertilizer use and agricultural yields. This counterintuitive finding led my students and I to ask a series of questions: What could possibly be causing this? Market dynamics? Changes in reporting? Overuse of other types of fertilizers? Changes in farming practices?

In short, we were asking "What's the mechanism?" That is, what are the details of the underlying interactions among farmers, markets, government agencies, subsidies, and crops that produce this effect? We wanted to go deeper into what philosophers of science refer to as the *causal mechanism* that produced the observed effects of fertilizer subsidies.

I want to emphasize that I'm not criticizing the methods or conclusions of this particular study. Indeed, from what I know of the study, it appears to be a very competent and effective use of the tools for causal analysis that are currently available to social scientists. My key concern here is not with the study's validity. Rather, my concern is with the type of reasoning that the study's findings elicit in readers and the extent to which the available tools support and inform that type of reasoning.

11.2 Key Ideas

The rest of this chapter will focus on several basic contentions exemplified by the case above:

- *Quantitative methods for causal inference have advanced tremendously in the past several decades* – Work in a variety of fields, including statistics, computer science, philosophy, and econometrics, have greatly expanded the concepts, methods, and tools available for causal inference. The study above used a difference-in-differences design (Shadish et al., 2001), one of a variety of techniques for causal inference whose use has expanded dramatically since the 1980s.

- *Current methods for inferring causal models provide unsatisfying answer for some types of questions* – Current methods allow us to reason about a type of high-level descriptive causality phrased in terms of random variables (e.g., agricultural yield, fertilizer expenditure, household size, age of household head), but not necessarily in terms of the underlying mechanisms that are responsible for the changes in those variables.

- *Satisfactory explanations of causation, even for non-experts, often requires inferences about mechanisms* – My students and I didn't know anything about fertilizer subsidies or crop production in Nepal, but we nonetheless immediately jumped to questions about the underlying causal mechanism. The analysis, while extremely interesting, raised as many questions as it answered, and there wasn't a direct line from the conclusions of the study to the answers we sought.

- *Current quantitative methods for causal inference do not directly support reasoning about possible mechanisms* – It is easy to assume that our questions could have been answered with additional data, and that such answers wouldn't have required a change in methods. However, as I argue, the methods themselves would still be insufficient even if more data had been available.

- *Several promising new methods are in development* – Several recent developments in machine learning point to promising additions and elaborations of the tools available for causal modeling. These are likely to provide substantially better support for reasoning about causal mechanisms in the future.

In the rest of this chapter, I pursue three basic directions. First, I discuss two fascinating but almost entirely unconnected areas of research about causal inference and causal mechanism, respectively. Second, I describe the current gap between these areas and the opportunity associated with connecting them. Finally, I provide some pointers to potential methods and recently developed techniques that may connect these two areas and that could enable a powerful set of new methods for understanding causal mechanism.

11.3 Quantitative Methods for Causal Inference

The first literature I want to briefly describe deals with quantitative methods for causal inference.[1] This is a large and exceptionally diverse literature, so it is impossible to adequately describe the array of available methods in a few pages. As a result, this is at best a sketch rather than any attempt at a comprehensive account.

One of the most important things to understand about this literature is that causal inference has been studied in many different communities, and the work in these communities is remarkably consistent despite being developed nearly independently. Among the best known communities are those associated with the potential outcomes framework (e.g., Rubin, 2005), causal graphical models (e.g., Pearl, 2000; Spirtes et al., 2000), and quasi-experimental design (e.g., Shadish et al., 2001). Additional work has been done in physics, epidemiology, biology, economics, and many other fields. These communities do not address identical questions, nor have they developed identical methods. The terminology and mathematical notation of these fields can be maddeningly inconsistent. If you look past the surface, however, the work from these different communities is almost entirely compatible, despite what some proponents of one or the other school of causal inference will tell you.

11.3.1 Progress

In the span of a few decades, these disparate research communities have developed conceptual frameworks and practical methods that have greatly advanced scientific research, particularly for fields in which experimentation is impractical, impossible, or unethical. I focus on two basic categories of work: methods for estimating a single causal dependence, and methods for learning the causal structure among a large set of variables.

Methods for estimating a single causal dependence are obviously useful. They answer the question: What will be the causal effect of a treatment X on an outcome Y? In some cases, this may be the treatment effect averaged over all values of a set of covariates, or it may be a treatment effect conditioned on a set of other covariates.

A key challenge of making such estimates accurately, of course, is that one or more of the covariates (Z), often referred to as confounders, may be a common cause of both treatment (X) and outcome (Y). Conditioning on a

---

[1] When I use the terms "causal inference" or "causal modeling," I am referring broadly to a set of techniques intended to infer causal dependence from observational and experimental data.

large number of possible confounders is a key challenge of causal inference, and a wide variety of methods have been developed for doing so. These run the gamut from simply partitioning data by the joint values of the covariates to constructing complex non-parametric models of $P(X|Z)$ and $P(Y|Z)$ to test whether conditioning on Z renders X and Y conditionally independent. Propensity score models (Rosenbaum and Rubin, 1983) and doubly robust estimators (Bang and Robins, 2005) fall toward the more complex end of this scale and are widely applied in social science and medicine. Finally, several quasi-experimental designs (Shadish et al., 2001) have been developed that, in some way, attempt to account for observed and unobserved confounders. These include Instrumental variable designs (Angrist et al., 1996), within-subject designs (Greenwald, 1976), interrupted time-series designs (Shadish et al., 2001), and regression discontinuity designs (Hahn et al., 2001).

Unsurprisingly, *time* is often a major factor in the reasoning that goes into using these methods. Temporal ordering is often used to rule out certain causal models as implausible. For example, a given treatment may be known to occur before a given outcome, thus ruling out one possible causal direction. In addition, investigators are often advised to confine conditioning operations to "pre-treatment" variables, a constraint that avoids the potential error of conditioning on a common *effect* of treatment and outcome, which can introduce bias into estimated treatment effects (Elwert and Winship, 2014).

In addition to methods for learning single causal dependencies, methods have been developed that can jointly estimate the causal dependencies among a large set of variables. One might imagine that such methods merely involve repeated application of methods for learning single dependencies, but such an approach would greatly underestimate what is possible by jointly considering causal dependence among sets of variables. First, knowledge of the dependence (or, rather, the conditional independence) between a pair of variables is highly informative for analyzing the possible causal dependence among other pairs with a given set of variables. Thus, joint learning algorithms perform far better than would be achieved by naive application of the simpler methods for learning single dependencies. Second, the jointly learned structure has many other desirable properties that have been formalized into the theory of causal graphical models (Pearl, 2000; Spirtes et al., 2000).

Causal graphical models (or, slightly more generally, directed graphical models) allow a generalized form of probabilistic inference that is extraordinarily powerful (Koller and Friedman, 2009; Pearl, 2014). First, given the values of any subset of variables in the model, algorithms exist that can accurately and tractably estimate the probability distribution of any other disjoint subset of variables. Second, given a causal graphical model, we can estimate how intervening on any variable affects the probability distribution on any other set of variables. Such interventional distributions are distinct from the observational distribution. Third, effective and well-understood algorithms exist for learning the structure and parameters of causal graphical models from data. These algorithms fall into several families, including constraint-based algorithms (e.g., PC and FCI (Spirtes et al., 2000)), score-based algorithms (e.g., GES (Chickering, 2002)), and hybrid algorithms (e.g., MMHC (Tsamardinos et al., 2006)).

Surprisingly, work in simple forms of causal graphical models rarely considers time explicitly. While an investigator could choose to limit the search space of an algorithm to models that respect a given temporal order, this is relatively rarely done. In fact, one of the remarkable properties of algorithms that infer the structure of causal graphical models from observational data is that they learn a causal (and, thus, temporal) order over a set of variables using only information about conditional independence.

This is only the briefest of summaries of a vast body of work on causal modeling. I have omitted mention of many other topics that have been examined in great technical depth, including identifiability (Pearl, 2000), transportability (Bareinboim and Pearl, 2013), selection bias (Elwert and Winship, 2014), and many others. However, even these two classes of work represent major advances in our understanding of causality over what was known less than 50 years ago.

11.3.2 How Do We Know These Methods Work?

The work described here certainly sounds impressive, but a key question is "Do these methods work?" That is, do these methods learn accurate causal models? Fortunately, there is substantial evidence that they do. This evidence comes in at least three varieties. First, there are proofs about some properties of these algorithms that suggest they will perform well. For example, several algorithms for learning the structure of a causal graphical models have been proven sound and complete in the sample limit. That is, given infinite data, they will converge to the true causal structure.

Second, there is evidence from extensive simulation studies. For example, in the study of algorithms for constructing causal graphical models, some simulations will hypothesize a given model, generate a finite data set from that model, apply an algorithm to that data set to learn a model, and then compare the structure and parameters of that learned model to the original generating model. Such studies allow researchers to vary sample size, model complexity, and other factors to see how they affect the accuracy of the learned model.

Finally, there is some evidence from empirical studies. In these studies, methods for learning causal models are applied to real-world data sets and then the learned models are compared to some sort of ground truth. In some cases, this ground truth has been derived from experiments on the same or related systems. In other cases, researchers rely on domain experts to assess ground truth.

11.3.3 Example: Postgres

Let me give you an example that demonstrates the capabilities of available methods for causal modeling (particularly those that construct causal graphical models), empirical evaluation methods, and some of the remaining questions that such methods leave unanswered. Several years ago, my students and I started investigating how to empirically evaluate current methods for constructing causal graphical models. Specifically, we looked for empirical systems (as opposed to simulators) in which we could gather observational data and perform experiments. This would allow us to construct causal models from the observational data, and then evaluate the causal inferences of those models by comparing them to the results of the experiments.

Most such systems have significant drawbacks. Some, such as simple physical systems, are not sufficiently complex to seriously evaluate methods for causal modeling. Others, such as social networks, violate some of the basic assumptions of the most well-developed methods. Still others, such as social or economic systems, make it logistically difficult or impractical to conduct experiments. Still others are not easily reproducible by other researchers.

Remarkably, however, we were able to find several systems that met our goals and yet avoided all of these problems. Furthermore, those systems were in our own backyard, so to speak. Specifically, we identified three *computational* systems that are very complex, yet amenable to both observation and experimentation. We gathered voluminous data from systematic experiments on these systems, and then we used the data to evaluate the effectiveness of algorithms that construct causal graphical models (Garant and Jensen, 2016).

I'll describe one of these systems, and the resulting evaluation, in more detail. PostgreSQL (or just Postgres) is a widely used open-source database management system. Postgres has been in continuous development for more than 30 years by over 400 developers, and it now consists of roughly 1.1 million lines of code.[2] Databases such as Postgres process queries expressed in a formal language (SQL) and return database records that match those queries. The precise manner in which a query is processed can be very complex, and

---

[2] See: https://www.postgresql.org/.

the decisions made by a modern database system depend on a variety of factors about the query, the data, indexing, and machine-specific factors. How a query is processed also depends on a set of configuration parameters whose values are typically determined by a database administrator. How to set those parameters for maximum performance remains a bit of a mystery, even to experienced database administrators, so a natural question is the causal effect of those settings on database performance.

We gathered experimental data in which a single row in the data corresponds to a given query, the database records that correspond to that query, and how that query is processed by Postgres. Treatments are the configuration parameters: the use of indexing, page access cost, and working memory allocation. Outcomes include query runtime, the number of blocks read from shared memory, temporary memory usage, and a fast memory cache. Additional covariates included aspects of the query itself (e.g., the number of joins, number of grouping operations, length), statistics of the referenced tables, and the number of rows retrieved by the query.

Our experiments consisted of installing a version of Postgres on a single machine, loading a large data set into the database, running over 11,000 queries under each of eight possible joint values of the three binary treatment variables, and recording data about each of the approximately 90,000 runs. The data set and queries were gathered from Stack Overflow, a large website about programming that makes their data publicly available. Neither the Postgres software, the Stack Overflow data set, nor the queries were written by us.

Given such an extensive set of experimental data, learning a causal model would be almost trivial. However, we wished to evaluate methods for learning causal models from observational data. Thus, we needed observational data in which treatment was not randomized and which had only one treatment configuration per query. To obtain such data, we used a non-random sampling procedure. For each query, the procedure selected one treatment configuration based on a probability distribution determined by the value of one of the covariates (number of rows retrieved, which is logically prior to treatment). This produces what can be considered the "moral equivalent" of observational data.

We then applied several analysis methods to the resulting data set. I'll discuss the results of two such methods. First, we applied a state-of-the-art method for learning a non-causal, associational model – a large non-parametric model called a random forest and known to be a high-quality conditional probability estimator. We also applied greedy equivalence search (GES), a well-known method for learning a causal graphical model.

The structure of the causal graphical model constructed by GES is shown in Figure 11.1. The variables in the figure are grouped into sets corresponding to

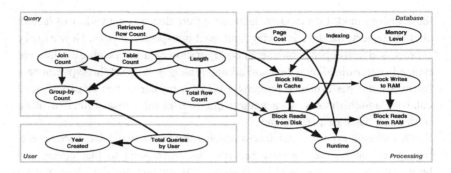

Figure 11.1 Causal graphical model for Postgres query processing.

the user, the query, the database configuration, and the query processing. This version of the model structure omits the weakest learned causal dependencies to focus on only the high-level structure of the model. This particular learned structure makes intuitive sense to a computer scientist: user variables only cause query variables, query variables only cause processing variables, and database configuration variables only cause processing variables. Note also that this learned model respects the expected temporal order, even though temporal ordering was not an input to the modeling process.

While the model appears to make intuitive sense, a more important question is whether the model's quantitative estimates of causal effect match the experimental estimates. Our quantitative evaluation says that they do. We show that the estimated distributions of interventional effect of the causal model learned by GES are far more accurate than the estimated distributions of the non-causal, associational model. Figure 11.2 provides one example of such distributions corresponding to a single set of values for the configuration variables and the distribution for one outcome variable.

11.3.4 Regrets

On a variety of levels, I am intrigued and pleased with these results. They demonstrate a somewhat unexpected correspondence between the model that an entirely algorithmic approach can construct and the high-level mental model that a relatively experienced database administrator might have. They also cleanly demonstrate an important underlying principle of causal inference – that intervention and observation are fundamentally different phenomena.

That said, however, there is something missing. As a computer scientist, I care about the underlying computational mechanisms that leads to the behavior described by the causal graphical model. Computer scientists care about

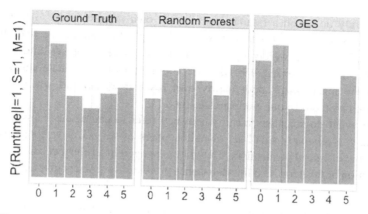

Figure 11.2 Actual (Ground Truth) and inferred (Random Forest and GES) distributions of query runtime given a specific intervention on database configuration variables.

how the database works at the level software, and how low-level changes in the actual implementation of the database would affect the runtime or hardware usage of a given combination of query and data. This is the level of understanding that matters most to me. And yet the learned structure and parameters of the causal graphical model, however impressive, tell me almost nothing of those details.

11.4 Causal Mechanism

This brings me to the second broad area that I want to describe, research into the nature of causal mechanism. As before, this will be a sketch, rather than a comprehensive account. Partially, this is because I am a newcomer to this literature, and I am not broadly knowledgeable about philosophy of science. In addition, I am not so much interested in the details of the definitions of causal mechanism, but more in the pre-conditions of a formal representation that allow causal mechanisms to be expressed. In machine learning, we are far enough away from being able to formally represent mechanism that the details of the definitions are not (yet) particularly important. Thus, I'll focus on the most basic elements of various definitions.

11.4.1 What Is a Mechanism?

Practicing scientists routinely discuss causal mechanism, either explicitly or implicitly in the form of research questions and hypotheses that they formulate.

I certainly do so on a regular basis in my own research, and some of my best work has been in cases in which I sought to identify the underlying mechanisms for a widely observed phenomenon (e.g., Jensen and Cohen, 2000; Oates and Jensen, 1999). I also routinely exhort graduate students to "find the mechanism" both in my own research group and when I teach research methods to our graduate students.

So, what do scientists mean when they refer to causal mechanism? Fortunately, a relatively small set of philosophers of science have long studied the topic of causal mechanism. They have engaged in a robust debate over the definition of this term, and this debate has produced several concise definitions:

> Mechanisms are entities and activities organized such that they are productive of regular changes from start or set-up to finish or termination conditions. (Machamer et al., 2000)

> A mechanism for a phenomenon consists of entities and activities organized in such a way that they are responsible for the phenomenon. (Illari and Williamson, 2012)

> A mechanism is a structure performing a function in virtue of its component parts, component operations, and their organization. (Bechtel and Abrahamsen, 2005)

These definitions of causal mechanism, deriving from the literature of the philosophy of science, contrast sharply with a much more narrow definition typically used in the literature of statistics, social science, and causal inference (e.g., Imai et al., 2011; VanderWeele, 2009). In those latter fields, "mechanism" often refers to the decomposition of a total causal effect into the effects from two or more causal pathways, sometimes with inferences about mediating variables. This notion of mechanism can be (and often is) directly represented in a causal graphical model, and primarily concerns the completeness of the set of variables and pathways expressed in that model. The definitions from statistics and from philosophy of science are not incompatible, but the latter set is far more expansive and evocative of potential future research directions, and we will use this more expansive definition for the remainder of this chapter.

11.4.2 Example: Surgical Therapies for Obesity

To better appreciate how philosophers of science often define causal mechanism, it is helpful look at another example. Last year, a study was published in the journal of the International Society for Microbial Ecology (Ilhan et al.,

2017). The study examined patients before and after having two different types of gastric surgery. Each type of surgery attempts to reduce obesity in treated patients, but the two types of surgery have substantially different outcomes. One type (referred to as Roux-en-Y gastric bypass or RYGB) tends to achieve greater weight loss than the other type (laparoscopic adjustable gastric banding or LAGB).

Both surgeries are intended to decrease the size of the stomach, but each imposes substantially different physical changes on the stomachs of treated patients. In RYGB, the stomach is divided into two portions, and the small intestine is rerouted to connect both portions. In LAGB, an adjustable band is placed around the top portion of the stomach. The most obvious causal mechanism that accounts for the differential effect of these surgeries is that the changes they create in the stomach change how food is processed and how patients feel when consuming food.

However, the study's authors provide evidence for an alternative conjecture: that the surgeries produce different changes in the microbiome of affected patients. Their study indicated that patients who received RYGB had a substantially more diverse stomach microbiome, and that such diversity is typical for non-obese patients who do not receive surgery, while patients who received LAGB had a low-diversity stomach microbiome, and that low diversity is typical for obese patients who receive surgery.

Again, what's important here is not the validity of this particular piece of research, but rather the questions that these findings raise about the underlying causal mechanism for obesity. Research into the human microbiome has created a variety of fundamentally new conjectures about the underlying causal mechanism behind many health effects, including this one. These results suggest that, perhaps, surgery could be entirely avoided if an alternative approach could be found to encourage a diverse microbiome.

11.4.3 Decomposing Mechanisms

Returning to the definitions, let's dig somewhat deeper. Each of the definitions above reference something like "entities" and "activities." Reviewing a set of definitions in the literature, Illari and Williamson (2012) note that:

> There is consensus that mechanistic explanation involves decomposition, and mechanisms have two distinct kinds of constituents. We have "entities", "parts" and "component parts" used for the bits and pieces of the mechanism, and "activities", "interactions" and "component operations" for what those bits and pieces do.

Table 11.1. *Entities and activities.*

System	Entities	Activities
Nepal farm subsidies	Farmers, Land, Crops, Subsidies	Decisionmaking, fertilizer application, crop growing
Database querying	Queries, Configurations	Query processing, table scans
Obesity therapy	Patients, Meals	Eating, digestion, metabolism

Craver and Darden (2013) note that:

> Mechanisms are composed of both entities and activities. The entities are the parts in the mechanism with their various properties. Activities are the things that the entities do; they are the producers of change.

and that "The entities have sizes, shapes, positions, and orientations. The activities have orders, rates, and durations."

To a computer scientist, this decomposition of causal mechanisms into entities and activities comes as no surprise. This sort of decomposition mirrors similar ones in computer science. For example, we often describe computer software as consisting of data structures ("entities") and algorithms ("activities"). Similarly, much of the work in artificial intelligence can be decomposed into a knowledge representation ("entities") and reasoning ("activities").

Efforts to reason about causal dependence can also be summarized by the concepts of entities and activities, even when the actual analysis conducted is far more narrow. Table 11.1, summarizes the entities and activities for the three examples outlined earlier.

It is also worth noting that nearly all descriptions of causal mechanisms are made using even lower-level mechanisms based on some more "fundamental" field. For example, a description of a biological mechanism might reference physical, chemical, or electrical phenomena that have, in turn, been described by earlier researchers as causal mechanisms in and of themselves.

Thus, mechanisms can be arranged hierarchically. This closely parallels ideas laid out by Brian Arthur in his magnificent and deceptively simple book *The Nature of Technology* (Arthur, 2009). Arthur notes the hierarchical fashion in which almost any technology combines other lower-level technologies. These hierarchies eventually ground out in fundamental phenomena based on "causal domains" such as mechanical, electrical, computational, psychological, and sociological phenomena. Table 11.2 summarizes the lower-level

Table 11.2. *Phenomena for causal mechanisms.*

System	Lower-level causal domains
Nepal farm subsidies	Sociological, economic, agricultural, botanic, administrative
Database querying	Computation, physical, informational, electronic
Obesity therapy	Biochemical, bacteriological, metabolic, psychological, physical

domains which ground the causal mechanisms in the various examples given to date.

Finally, note the importance of *time* to these descriptions of causal mechanism. The central role of "activities" implies a causal process that unfolds over time ("The activities have orders, rates, and durations."). It also implies the potential for interacting, concurrent, and ongoing causal processes, as well as processes that proceed at different rates depending on the conditions under which the causal process executes.

11.4.4 What Should Mechanisms Do?

These definitions may help us describe what a causal mechanism is, but a more important question is what a useful model of a causal mechanism does. Such an operational definition could help distinguish between a superficial account and a high-quality model of a causal mechanism. Craver and Darden (2013) outline three remarkably commonsense tests to do just this. They phrase their tests in terms of two questions and one directive:

- *"And how does that work?"* – A black box with inputs and outputs is not a sufficient account of causal mechanism. Instead, a deep understanding of mechanism should facilitate a plausible description of the inner workings of that box. For example, merely stating that fertilizer subsidies in one region of Nepal decrease the use of chemical fertilizers is insufficient. Instead, we want an account of the details of the interactions among economic decisions, farming behavior, and other factors that actually lead to the effect.
- *"What if that worked differently?"* – Superficial knowledge of the causal connections among a set of factors (e.g., A causes B, and B causes C) is not a sufficient account of causal mechanism. Instead, understanding the

mechanism should allow us to answer the question of what would occur if D were substituted for B. For example, an understanding of mechanism should allow us to infer what might occur if the structure of the economic subsidy, farming practices, or specific crops had been different.

- *"Build it"* – Perhaps the most stringent test of a causal mechanism is whether it contains sufficient detail to serve as a blueprint to construct an actual system that would work similarly. For example, given an account of the causal mechanism operating in the fertilizer subsidy example, could we construct a farming community and government program that would reliably exhibit the same behavior?

These questions emphasize the depth of knowledge and understanding about inner workings that characterize a high-quality account of a causal mechanism. Craver and Darden (2013) contrast this with more shallow accounts, saying that: "The distinguishing mark of a superficial, phenomenal model is that it describes the behavior of the mechanism without describing how the mechanism works."

Connecting back to Arthur's (2009) account of technologies, we can add another question that helps distinguish shallow from deep accounts of causal mechanism: "Can it make use of known mechanisms in more fundamental causal domains?" For example, we would expect that a deep account of a causal mechanism for the fertilizer example would reference known mechanisms in fields such as sociology, economics, agriculture, botany, and political science. Knowing those lower-level mechanisms would provide ready-made components for our higher-level model of mechanism.

11.5 The Gap

Juxtaposing these two sketches, of quantitative methods for causal inference on the one hand and causal mechanism on the other, leads to an obvious question: To what extent do the current suite of quantitative methods aid in the discovery of satisfying accounts of causal mechanisms? If, like philosophers of science and many practicing scientists, we believe that causal mechanisms are a fundamental goal of scientific research, we would hope that existing tools for causal modeling would directly aid the search for such mechanisms.

11.5.1 What's Present and What's Missing?

There are ways in which existing methods succeed. If a researcher has a clear mechanism in mind, that mechanism often implies one or more analyses that can be conducted using existing quantitative methods. Indeed, one mark of a capable sociologist, epidemiologist, or biologist is their ability to translate a hypothesized causal mechanism into the framework of potential outcomes or causal graphical models. These frameworks typically require reducing the operation of a complex mechanism into a data set in which rows represent independent, identically distributed data instances and columns represent random variables. The resulting analysis can be viewed as a test of one or more hypotheses about the mechanism, including the overall hypothesis that the proposed mechanism is valid or some more specific hypothesis that assumes the mechanism's validity.

However, when we apply these quantitative methods, one output is often what we refer to as a "causal model." An entirely plausible question is the extent to which such a model is a satisfactory representation of a causal mechanism. In very important ways, they are not. Recall Craver and Darden's (2013) comment that "The distinguishing mark of a superficial, phenomenal model is that it describes the behavior of the mechanism without describing how the mechanism works." Both the frameworks of potential outcomes and causal graphical models produce models that describe the effects of the mechanism in terms of changes in the values of random variables rather than describing "how the mechanism works." They provide only extremely narrow help in answering the question "What if it worked differently?" and they provide almost no assistance in fulfilling the directive to "Build it."

It is easiest to see what's missing in the case of the potential outcomes framework. In a typical analysis, a regression model is used to estimate the degree to which a single treatment causes a single outcome in the context of a set of possible confounders. Recall Illari and Williamson's (2012) note that mechanisms have two distinct kinds of constituents: entities ("'parts' and 'component parts' used for the bits and pieces of the mechanism") and activities ("'interactions' and 'component operations' for what those bits and pieces do."). Regression models typically and implicitly represent only a single type of entity corresponding to what is sometimes called the "unit of analysis" or "outcome" of a sampling process. In the case of the fertilizer subsidy analysis, this corresponds to a single farmer (and their associated household and agricultural land). Regression models typically also represent very limited types of activities. They cannot easily represent a variety of common phenomena in mechanisms, including interactions among entities, the creation and destruction of entities, and changes in the structure of a system or function of component entities over time.

Attempts have certainly been made to extend the expressiveness of regression models, including work in multi-level models that can represent multiple types of entities, structural equation models that can represent some types of feedback effects, and models for estimating effects in data sets in which instances are statistically dependent. As I describe later in this chapter, the ideas behind these methods point toward a more comprehensive solution. However, even these partial solutions remain largely outside the mainstream of use by social scientists, physical scientists, medical researchers, and others.

The causal graphical models framework provides a slightly better representation for causal mechanisms, but there is still a large gap between what these models represent and what scientists conceive of as a causal mechanism. As outlined previously, causal graphical models elegantly represent how large numbers of causal dependencies interact, and they allow inference about the probability distributions of any set of variables in the model given observations of the values of any other set. Computer scientists, statisticians, and others have developed multiple well-understood methods for estimating the structure and parameters of such models from data, and these methods can incorporate a variety of prior knowledge about the structure and parameters. Even these capabilities, however, leave a large gap between what causal graphical models can represent and what scientists typically call a causal mechanism. I am not the first to note this gap, for example see Aalen et al. (2016). As with regression models, causal graphical models typically represent only a single type of "entity" and cannot easily represent many types of "activities" that appear common to complex causal mechanisms. Also like regression models, attempts have been made to extend their expressiveness, and I outline these in more detail later in this chapter.

In the discussion of causal mechanisms, I outlined the three tests suggested by Craver and Darden (2013), and I suggested adding an additional test: asking whether the account can incorporate known mechanisms from more fundamental causal domains. For example, can an account of the causal mechanism of fertilizer subsidies incorporate previously known lower-level mechanisms about economic effects of subsidies on prices, biological effects of fertilizers on crop production, or sociological effects of government subsidies given to only some members of a community? This question highlights the gap between current quantitative methods and robust models of causal mechanisms. Neither regression models nor causal graphical models, as they are typically used, allow researchers to incorporate existing models of lower-level domains in this way.

To summarize, researchers who can already describe a causal mechanism can sometimes use existing quantitative methods to explore the validity of a single such hypothesis, but the methods do little to aid discovery of, and reasoning about, causal mechanisms. Knowing the mechanism allows construction of an analysis, but the results of that analysis do relatively little to inform what alternatives would better model the mechanism.

11.5.2 An Historical Note

By now, my larger point is probably becoming clear: I am advocating for a transition from a relatively simple set of analytical methods (that describe the effects of causal mechanisms) to a more expressive and capable set of methods (that describe the functioning of causal mechanisms). Such a transition is hardly unprecedented. In his groundbreaking book on causal inference, Judea Pearl advocated strongly for an analogous transition, from merely analyzing statistical association to analyzing causal dependence. He said: "I see no greater impediment to scientific progress than the prevailing practice of focusing all of our mathematical resources on probabilistic and statistical inferences while leaving causal considerations to the mercy of intuition and good judgment" (2000, xiii). The form of my argument is similar: Why leave inferences about causal mechanism to the mercy of intuition and good judgment?

Pearl correctly criticizes prior statistical frameworks for having no explicit representation of causation, and he notes how the lack of such a representation impedes analytic reasoning about causation. However, using impoverished representations of causal mechanisms means that current quantitative methods run a similar risk. Causal mechanisms have the same relation to patterns of observed causal dependence as current models of causal dependence have to patterns of observed statistical dependence. Just as patterns of statistical dependence describe the effects of causal dependence, patterns of causal dependence describe the behavior of causal mechanisms. Just as causal graphical models enable more effective reasoning about interventions than do patterns of statistical dependence, models of mechanisms would allow more effective reasoning about interventions than do causal graphical models.

What opportunities are we currently missing? First, limiting our reasoning about causal mechanisms directly limits the set of interventions we can consider. Existing causal models are impressive partially because they allow counterfactual reasoning (reasoning about a joint distribution of data that is different than the one observed in the training data). However, existing causal models are completely silent about the potential effects of actions that have never before been fully observed. In contrast, knowledge of a causal mechanism can imply the existence of potentially successful interventions that have never been fully observed. For example, in the example of various gastric surgeries, the proposed mechanism for the success of gastric bypass surgery (change in the diversity of the stomach microbiome) implies that other approaches for manipulating microbiome diversity could have dramatic effects on weight loss. This is true even if the weight-loss effects of those approaches has never been measured.

Second, limiting our reasoning about causal mechanisms denies us the ability to use important classes of evidence for inferring causal mechanisms. For example, reasoning about causal mechanisms can allow previously known lower-level causal mechanisms to provide constraints on possible higher-level mechanisms (e.g., "that's just not physically possible…"). Existing causal models, expressed in the languages of regression equations or causal graphical models, only very rarely allow this.

11.5.3 Why Does This Gap Exist?

Criticism is easy. Merely noting that existing methods are inadequate for some important purpose does not imply that better methods can and should be developed. At the very least, we need to ask why existing methods are the way they are, why improved methods have not yet been developed, and whether there exist insurmountable barriers to improvement.

Why are existing methods the way they are? One reason seems relatively clear: simplicity. All models expressed as regression equations or causal graphical models have the same formal semantics, regardless of the behavior of the causal mechanism they are designed to represent, and this formal semantics has proven to be highly flexible. Because of this flexibility, current methods can represent and analyze an extremely wide range of phenomena. This flexibility is a source of power, but also a weakness. The current languages for causal inference almost never match the language that a practicing scientist would naturally use to represent a hypothesized mechanism. Instead, practicing scientists are required to translate (some might say "shoehorn") their knowledge of mechanism into a single language of causal inference. Some entities and processes cannot be represented effectively, and most constraints and knowledge from lower-level causal mechanisms cannot be represented easily and explicitly within the model.

Why haven't improved methods been developed? Truthfully, many improvements have been developed, but they only satisfy some of the properties necessary to represent causal mechanisms, and adoption has been slow. One reason for this is that, until fairly recently, the computing technologies and expertise necessary to use these methods have not been widely available. Fortunately, this is changing. Many researchers now have routine access to large-scale computational resources, research communities are developing shared resources such as online data repositories and specialized ontologies, and many fields are developing cadres of researchers with in-depth knowledge of computing (e.g., bioinformatics and computational social science).

Do insurmountable barriers exist to developing better methods? Almost certainly not. As I will describe in the next section, recent progress in machine learning, causal inference, statistics, and programming languages suggest that much more expressive classes of models are currently in development, and these could provide dramatic increases in our ability to represent and reason about causal mechanisms. Even more intriguingly, there appears to be a confluence of multiple areas of work on some common concepts and technologies.

11.6 New Directions

Given that such a gap exists, what should we be doing to bridge it? I would be delighted if I could provide a ready-built solution or even an effective blueprint. Unfortunately, I cannot. What I can provide is an overall direction suggested by recent developments in artificial intelligence and machine learning. First, I summarize some recent developments in causal graphical models, and then I provide a sketch of how a new approach – probabilistic programming – could provide a rich basis for representing causal mechanism.

A high-level summary of these alternatives is shown in Figure 11.3. The y-axis of the figure represents the expressiveness of a given language for causal modeling. The x-axis represents the degree to which models in that language can be estimated from data. Commonly used methods for linear regression and causal graphical models are represented near the bottom of the space. Higher up the scale of expressiveness are forms of graphical models that can express relational and temporal dependence. Higher still are traditional simulation languages, although very little about these models can be estimated directly from data. Finally, probabilistic programs are shown near the top of the figure, with an indication that their ability to be estimated from data is under rapid development.

11.6.1 Extended Forms of Graphical Models

Given the capabilities of causal graphical models, and their wide use within the research community for representing causal dependence, it seems reasonable to ask whether their capabilities can be extended to provide better models of causal mechanisms. The short answer is yes. Various extensions to directed graphical models have been developed and analyzed since the year 2000. Remarkably, these extensions correspond to almost precisely the components that philosophers of science have said are necessary to represent causal mechanisms: entities and activities.

In terms of entities, several researchers have developed extensions to directed graphical models, often referred to as "relational models," that represent the interactions among multiple entities (Getoor and Taskar, 2007) and that model the creation and destruction of both entities and the relationships between entities (Getoor et al., 2001). More recently, researchers have extended these models to have an explicitly causal semantics, extended the theory that links that causal semantics with observable statistical properties, and applied that theory to show how such models can be learned from observational data (Lee and Honavar, 2016a, 2016b, Maier, 2014; Maier et al., 2013).

In terms of activities, researchers have also developed extensions that explicitly consider time, often referred to as "dynamic models," that represent how systems of variables change over time (a plausible representation of an "activity"), including both discrete-time (Ghahramani, 1998; Murphy, 2002) and continuous-time models (Nodelman et al., 2002). Very recently, researchers have developed models that merge both relational and temporal elements simultaneously, and they have shown how their structure and parameters can be learned from observational data (Marazopoulou et al., 2015).

11.6.2 Probabilistic programs

This proliferation of different model formalisms, including many types of models not mentioned in this chapter so far, has led researchers in artificial intelligence and machine learning to consider a larger and more general class of probabilistic models. This class of models is represented in the form of a programming language, and a given model is a program in that language (Gordon et al., 2014; Pfeffer, 2016; van de Meent et al., 2018). Such probabilistic programs are written in Turing-complete programming languages that have syntax and semantics such that programs can be thought of as a representation of a joint probability distribution of the variables in the program, and program executions can be thought of as representing samples from that distribution.

Probabilistic programming languages can be characterized as having three key features: (1) Language primitives that allow for sampling and thus define

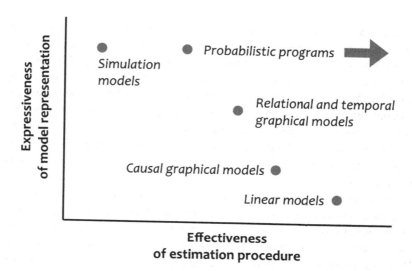

Figure 11.3 Comparison of alternative technologies for representing causal mechanisms.

simple probabilistic dependence (e.g., $y = Normal(x, 1)$, which says that the variable y is normally distributed with mean x and variance 1); (2) Language primitives that allow assertion and thus provide conditioning (e.g., $y > 2$, which limits the joint distribution represented by the program to only those executions in which $y > 2$); and (3) Inference algorithms that allow tractable estimation of the joint distribution associated with any given program.

To date, many probabilistic programming languages have been developed, including Figaro (Pfeffer, 2016), Church (Goodman et al., 2012), Venture (Mansinghka et al., 2014), Anglican (Tolpin et al., 2015), Stan (Carpenter et al., 2017), and many others. This field is still in a very early stage, and the promise of efficient, generalized inference has yet to be realized. Still, the expressiveness of these languages offers extraordinary power, and many researchers in AI and machine learning foresee that probabilistic programming will be a foundational technology for a broad range of future AI systems.

What is attractive about probabilistic programming for representing and reasoning about causal mechanism? There are several immediate answers. First, many programming languages have a clear causal semantics that makes them a natural choice for representing and reasoning about causal dependence. Second, many programming languages allow explicit representation of entities (e.g., objects) and activities (e.g., functions or methods), the key building blocks of causal mechanisms. Third, programming languages immediately provide ways of representing some formerly problematic features of causal mechanisms. For example, a formal property of all Turing-complete languages is conditional branching. In probabilistic programs, conditional branching provides for probabilistic dependence conditioned on a covariate value, an issue that has long bedeviled directed graphical models.[3] Finally, programming languages naturally allow for a hierarchical structure that appears to be an essential property of effective accounts of causal mechanisms.

Now, merely because probabilistic programming languages can express all of these things does not mean that language developers will be able to deliver on the promise of generalized inference for arbitrary programs. Nor does this mean that learning full programs from data will be feasible in the near future. However, probabilistic programs as a model representation appear far more promising than any prior formalism.

Indeed, some of these properties of probabilistic programming languages have already been exploited to represent a few complex generative (causal) processes in ways that are similar to the approach proposed here. For example, Lake et al. (2015) demonstrate how a probabilistic program can be used to model the process by which humans render handwritten characters. This same program is then used to model the variability of handwriting for previously unseen characters. While the goal was not to infer the effects of interventions in a causal process, the learned program could clearly have been used for that purpose and probably would have been highly effective.

Another key idea is that, while probabilistic programming languages are perhaps the closest thing we have to an ideal representation for causal mechanisms, even they are not necessarily the end point. Even probabilistic programs are less expressive than we might want, because (in their most raw form) they most naturally express only one type of mechanism: computer programs. It seems reasonably clear that, using Turing complete programming languages, we can construct domain-specific languages that elegantly describe mechanisms in other domains. However, it remains to be seen whether general approaches can be developed for learning programs in such domain-specific languages.

---

[3] For example, the if statement in a typical probabilistic programming language provides a simple method to model a population consisting of two or more subgroups with very different (and potentially incompatible) causal models.

11.7 Conclusions

I hope that my efforts to pull together this set of disparate ideas has at least been informative. I particularly hope that it encourages greater interest in reaching beyond each of our narrow subfields to make connections with other fields. I was surprised and delighted when I first saw the rich connections between my own work on expanding the expressiveness of causal graphical models and the existing scholarship on causal mechanisms, and I hope I have been able to convey some of the enthusiasm and delight that this connection has produced for me.

Acknowledgments

Dan Garant and Amanda Gentzel contributed greatly to my thinking on empirical evaluation of algorithms for causal modeling. Marc Maier and Katerina Marazopoulou contributed greatly to my thinking about how to extend causal graphical models to express relational and temporal dependence. Finally, thanks to Vikash Mansingka, Josh Tenenbaum, and Andrew McCallum for introducing me to the power and promise of probabilistic programming.

References

Aalen, Odd Olai, Røysland, Kjetil, Gran, Jon Michael, Kouyos, Roger, and Lange, Tanja. 2016. Can we believe the DAGs? A comment on the relationship between causal DAGs and mechanisms. *Statistical Methods in Medical Research*, 25(5), 2294–2314.

Ahn, Woo-Kyoung, Kalish, Charles W., Medin, Douglas L., and Gelman, Susan A. 1995. The role of covariation versus mechanism information in causal attribution. *Cognition*, 54(3), 299–352.

Albert, David Z. 2000. *Time and Chance*. Cambridge, MA: Harvard University Press.

Allison, Henry E. 2004. *Kant's Transcendental Idealism*. 2nd edn. New Haven, CT: Yale University Press.

Alston, William P. 1998. Empiricism. In *Routledge Encyclopedia of Philosophy*. Taylor and Francis. https://www.rep.routledge.com/articles/thematic/empiricism/v-1.

Angrist, Joshua D., Imbens, Guido W., and Rubin, Donald B. 1996. Identification of causal effects using instrumental variables. *Journal of the American Statistical Association*, 91(434), 444–455.

Anscombe, Gertrude Elizabeth Margaret. 1971. *Causality and Determination*. Cambridge: Cambridge University Press.

Arntzenius, Frank, and Greaves, Hilary. 2009. Time reversal in classical electromagnetism. *British Journal for the Philosophy of Science*, 60(3), 557–584.

Arthur, W. Brian. 2009. *The Nature of Technology: What it Is and How it Evolves*. New York: Simon and Schuster.

Au, Yin Chung. 2016. Synthesising heterogeneity: trends of visuality in biological sciences circa 1970s - 2000s. PhD thesis, University College London.

Bacciagaluppi, Guido. 2006. Probability, arrow of time and decoherence. *Studies in History and Philosophy of Science Part B: Studies in History and Philosophy of Modern Physics*, 38(2), 439–456.

Bang, Heejung, and Robins, James M. 2005. Doubly robust estimation in missing data and causal inference models. *Biometrics*, 61(4), 962–973.

Bareinboim, Elias, and Pearl, Judea. 2013. A general algorithm for deciding transportability of experimental results. *Journal of Causal Inference*, 1(1), 107–134.

236

Barndorff-Nielsen, Ole E., and Shephard, Neil. 2001. Non-Gaussian Ornstein–Uhlenbeck-based models and some of their uses in financial economics. *Journal of the Royal Statistical Society: Series B (Statistical Methodology)*, **63**(2), 167–241.

Barnes, Jonathan. 2013. *The Presocratic Philosophers*. New York: Routledge.

Bechlivanidis, Christos, and Lagnado, David A. 2013. Does the "why" tell us the "when"? *Psychological Science*, **24**(8), 1563–1572.

— 2016. Time reordered: Causal perception guides the interpretation of temporal order. *Cognition*, **146**, 58–66.

Bechtel, William. 2006. *Discovering Cell Mechanisms: The Creation of Modern Cell Biology*. Cambridge: Cambridge University Press.

— 2015. Can mechanistic explanation be reconciled with scale-free constitution and dynamics? *Studies in History and Philosophy of Science Part C: Studies in History and Philosophy of Biological and Biomedical Sciences*, **53**, 84–93.

Bechtel, William, and Abrahamsen, Adele. 2005. Explanation: A mechanist alternative. *Studies in History and Philosophy of Science Part C: Studies in History and Philosophy of Biological and Biomedical Sciences*, **36**(2), 421–441.

Bechtel, William, Burnston, Daniel, Sheredos, Benjamin, and Abrahamsen, Adele. 2014. Representing time in scientific diagrams. In: *Proceedings of the 36th Annual Meeting of the Cognitive Science Society*. Cognitive Science Society.

Beebee, Helen, Hitchcock, Christopher Read, and Menzies, Peter (eds). 2009. *The Oxford Handbook of Causation*. Oxford: Oxford University Press.

Beilock, Sian L., Gunderson, Elizabeth A., Ramirez, Gerardo, and Levine, Susan C. 2010. Female teachers' math anxiety affects girls' math achievement. *Proceedings of the National Academy of Sciences of the United States of America*, **107**(5), 1860–1863.

Ben-Yami, Hanoch. 2007. The impossibility of backwards causation. *The Philosophical Quarterly*, **57**(228), 439–455.

— 2010. Backwards causation still impossible. *Analysis*, **70**(1), 89–92.

Benoit, Kenneth. 2006. Duverger's law and the study of electoral systems. *French Politics*, **4**(1), 69–83.

— 2007. Electoral laws as political consequences: Explaining the origins and change of electoral institutions. *Annual Review of Political Science*, **10**, 363–390.

Bequette, B. Wayne. 2010. Continuous glucose monitoring: Real-time algorithms for calibration, filtering, and alarms. *Journal of Diabetes Science and Technology*, **4**(2), 404–418.

Black, Max. 1956. Why cannot an effect precede its cause? *Analysis*, **16**(3), 49–58.

Blackwell, Kenneth. 1989. Portrait of a philosopher of science. In Savage, C. Wade, and Anderson, C. Anthony (eds), Minnesota Studies in the Philosophy of Science, vol. 12. University of Minnesota Press, pp. 281–293.

Blais, André, and Carty, R. K. 1991. The psychological impact of electoral laws: Measuring Duverger's elusive factor. *British Journal of Political Science*, **21**(1), 79–93.

Bohm, David (ed). 1951. *Quantum Theory*. New York: Prentice-Hall.

Bramley, Neil R. 2017. Constructing the world: Active causal learning in cognition. Ph.D. thesis, University College London.

Bramley, Neil R., Gerstenberg, Tobias, and Lagnado, David A. 2014. The order of things: Inferring causal structure from temporal patterns. *Proceedings of the 36th Annual Meeting of the Cognitive Science Society*. Austin, TX: Cognitive Science Society, pp. 236–242.

Bramley, Neil R., Lagnado, David A., and Speekenbrink, M. 2015. Conservative forgetful scholars: How people learn causal structure through interventions. *Journal of Experimental Psychology: Learning, Memory & Cognition*, **41**(3), 708–731.

Bramley, Neil R., Mayrhofer, R., Gerstenberg, Tobias, and Lagnado, David A. 2017a. Causal learning from interventions and dynamics in continuous time. In: *Proceedings of the 39th Annual Meeting of the Cognitive Science Society*. Austin, TX: Cognitive Science Society.

Bramley, Neil R., Dayan, Peter, Griffiths, Thomas L., and Lagnado, David A. 2017b. Formalizing Neurath's ship: Approximate algorithms for online causal learning. *Psychological Review*, **124**(3), 301–338.

Bramley, Neil R., Gerstenberg, Tobias, Tenenbaum, Joshua B., and Gureckis, Todd M. 2018a. Intuitive experimentation in the physical world. *Cognitive Psychology*, **105**, 9–38.

Bramley, Neil R., Gerstenberg, Tobias, Mayrhofer, Ralf, and Lagnado, David A. 2018b. Time in causal structure learning. *Journal of Experimental Psychology: Learning, Memory & Cognition*, **44**(12), 1880–1910.

Bressler, Steven L., and Seth, Anil K. 2011. Wiener-Granger causality: A well established methodology. *Neuroimage*, **58**(2), 323–329.

Bruner, Jerome S., Jolly, Alison, and Sylva, Kathy (eds). 1976. *Play: Its Role in Development and Evolution*. Harmondsworth: Penguin.

Buehner, Marc J., and Humphreys, Gruffydd R. 2009. Causal binding of actions to their effects. *Psychological Science*, **20**(10), 1221–1228.

Buehner, Marc J., and May, Jon. 2002. Knowledge mediates the timeframe of covariation assessment in human causal induction. *Thinking & Reasoning*, **8**(4), 269–295.

— 2003. Rethinking temporal contiguity and the judgement of causality: Effects of prior knowledge, experience, and reinforcement procedure. *The Quarterly Journal of Experimental Psychology Section A*, **56**(5), 865–890.

— 2004. Abolishing the effect of reinforcement delay on human causal learning. *Quarterly Journal of Experimental Psychology Section B*, **57**(2), 179–191.

Buehner, Marc J., and McGregor, Stuart. 2006. Temporal delays can facilitate causal attribution: Towards a general timeframe bias in causal induction. *Thinking & Reasoning*, **12**(4), 353–378.

Burns, Patrick, and McCormack, Teresa. 2009. Temporal information and children's and adults' causal inferences. *Thinking & Reasoning*, **15**(2), 167–196.

Caplan, Ben, and Matheson, Carl. 2004. Can a musical work be created? *British Journal of Aesthetics*, **44**(2), 113–134.

Carpenter, Bob, Gelman, Andrew, Hoffman, Matthew D., Lee, Daniel, Goodrich, Ben, Betancourt, Michael, Brubaker, Marcus, Guo, Jiqiang, Li, Peter, and Riddell, Allen. 2017. Stan: A probabilistic programming language. *Journal of Statistical Software*, **76**(1), 1–32.

Carroll, Sean. 2010. *From Eternity to Here: The Quest for the Ultimate Theory of Time.* New York: Dutton.

Cartwright, Nancy. 1983. Causal laws and effective strategies. In: *How the Laws of Physics Lie.* Oxford: Oxford University Press, pp. 20–42.

— 1989. *Nature's Capacities and their Measurement.* Oxford: Clarendon Press.

Cheng, Patricia W. 1997. From covariation to causation: A causal power theory. *Psychological Review*, **104**(2), 367–405.

Cheng, Patricia W., and Novick, Laura R. 1990. A probabilistic contrast model of causal induction. *Journal of Personality and Social Psychology*, **58**(4), 545.

Chickering, David Maxwell. 2002. Optimal structure identification with greedy search. *Journal of Machine Learning Research*, **3**(Nov), 507–554.

Clark, Andy. 2013. Whatever next? Predictive brains, situated agents, and the future of cognitive science. *Behavioral and Brain Sciences*, **36**(3), 181–204.

Cobelli, Claudio, Schiavon, Michele, Dalla Man, Chiara, Basu, Ananda, and Basu, Rita. 2016. Interstitial fluid glucose is not just a shifted-in-time but a distorted mirror of blood glucose: Insight from an in silico study. *Diabetes Technology & Therapeutics*, **18**(8), 505–511.

Coenen, Anna, Rehder, Robert, and Gureckis, Todd M. 2015. Strategies to intervene on causal systems are adaptively selected. *Cognitive Psychology*, **79**, 102–133.

Coenen, Anna, Nelson, Jonathan D., and Gureckis, Todd M. 2017a. Asking the right questions about human inquiry. *Retrieved from psyarxiv. com/h457v.*

Coenen, Anna, Bramley, Neil R., Ruggeri, Azzurra, and Gureckis, Todd M. 2017b. Beliefs about sparsity affect causal experimentation. In: *Proceedings of the 39th Annual Conference of the Cognitive Science Society.* Austin, TX: Cognitive Science Society.

Coleman, James. 1990. *Foundations of Social Theory.* Cambridge, MA: Harvard University Press.

Colyvan, Mark. 1998. Can the eleatic principle be justified? *Canadian Journal of Philosophy*, **28**(3), 313–336.

Contessa, Gabriele. 2007. Scientific representation, interpretation, and surrogative reasoning. *Philosophy of Science*, **74**(1), 48–68.

Cook, Roy T. 2012. Impure sets are not located: A Fregean argument. *Thought: A Journal of Philosophy*, **1**(3), 219–229.

Cowling, Sam. 2014. No simples, no gunk, no nothing. *Pacific Philosophical Quarterly*, **95**(2), 246–260.

Cox, David Roxbee, and Isham, Valerie. 1980. *Point processes.* Monographs on statistics and applied probability, vol. 12. Boca Raton, FL: Chapman & Hall/CRC.

Cramer, John G. 1980. Generalized absorber theory and the Einstein-Podolsky-Rosen paradox. *Physical Review D*, **22**(2) 362–76.

Craver, Carl F., and Darden, Lindley. 2013. *In Search of Mechanisms: Discoveries Across the Life Sciences.* Chicago: University of Chicago Press.

Dash, Denver. 2003. Caveats for causal reasoning with equilibrium models. Ph.D. thesis, University of Pittsburgh.

Davidson, Donald. 1967. Causal relations. *The Journal of Philosophy*, **64**(21), 691–703.

Davis, James A. 1985. *The Logic of Causal Order.* Thousand Oaks, CA: Sage Publications.

Davis, Zachary, Bramley, Neil R., and Rehder, Robert E. 2018a. Causal learning from continuous variables in continuous time. In: *Proceedings of the 40th Annual Meeting of the Cognitive Science Society*. Austin, TX: Cognitive Science Society.

Davis, Zachary, Bramley, Neil R., Gureckis, Todd M., and Rehder, Robert E. 2018b. A causal model approach to dynamic control. In: *Proceedings of the 40th Annual Meeting of the Cognitive Science Society*. Austin, TX: Cognitive Science Society.

Dayan, Peter. 1993. Improving generalization for temporal difference learning: The successor representation. *Neural Computation*, 5(4), 613–624.

De Bal, Inge. 2017. Physical causal knowledge: A user's manual. Revisiting the debate about causation in physics from a use-perspective. Ph.D. thesis, Ghent University.

de Souza, Daniel Câmara, and De Luca, Jayme. 2015. Solutions of the Wheeler-Feynman equations with discontinuous velocities. *Chaos: An Interdisciplinary Journal of Nonlinear Science*, 25(1), 013102–11.

Dean, Thomas, and Kanazawa, Keiji. 1989. A model for reasoning about persistence and causation. *Computational Intelligence*, 5(2), 142–150.

Demirci, Kadir, Akgönül, Mehmet, and Akpinar, Abdullah. 2015. Relationship of smartphone use severity with sleep quality, depression, and anxiety in university students. *Journal of Behavioral Addictions*, 4(2), 85–92.

Deverett, Ben, and Kemp, Charles. 2012. Learning deterministic causal networks from observational data. In: *Proceedings of the 34th Annual Meeting of the Cognitive Science Society*. Austin, TX: Cognitive Science Society.

Di Blasi, Zelda, Harkness, Elaine, Ernst, Edzard, Georgiou, Amanda, and Kleijnen, Jos. 2001. Influence of context effects on health outcomes: a systematic review. *The Lancet*, 357(9258), 757–762.

Dowe, Phil. 1992. Wesley Salmon's process theory of causality and the conserved quantity theory. *Philosophy of Science*, 59(2), 195–216.

— 2000. *Physical Causation*. Cambridge: Cambridge University Press.

Dowe, Phil. 2008. Causal processes. *The Stanford Encyclopedia of Philosophy*.

Ducheyne, Steffen. 2011. Newton on action at a distance and the cause of gravity. *Studies in History and Philosophy of Science*, 42(1), 154–159.

Dummett, A.E., and Flew, Antony. 1954. Symposium: Can an effect precede its cause? *Proceedings of the Aristotelian Society, Supplementary Volumes*, 28, 27–62.

Dummett, Michael. 1964. Bringing about the past. *The Philosophical Review*, 73(3), 338–359.

Dupré, John. 1993. *The Disorder of Things*. Cambridge, MA, and London: Harvard University Press.

Duverger, Maurice. 1951. *Les partis politiques*. Paris: Librairie Armand Colin.

— 1959. *Political Parties: Their Organization and Activity in the Modern State (second English revised edition)*. London: Methuen & Co.

— 1961. *Les partis politiques (quatrième edition)*. Paris: Librairie Armand Colin.

Eames, Elizabeth R. 1989. Cause in the later Russell. In: Savage, C. Wade, and Anderson, C. Anthony (eds), *Minesota Studies in the Philosophy of Science*, vol. 12. University of Minnesota Press, pp. 264–280.

Earman, John. 1972. Notes on the causal theory of time. *Synthese*, 24(1–2), 74–86.

— 1976. Causation: a matter of life and death. *Journal of Philosophy*, 73(1), 5–25.

Earman, John, Smeenk, Christopher, and Wüthrich, Christian. 2009. Do the laws of physics forbid the operation of time machines? *Synthese*, **169**(1), 91–124.

Eells, Ellery. 1991. *Probabilistic Causality*. Cambridge: Cambridge University Press.

Effingham, Nikk. 2012. Impure sets may be located: a reply to Cook. *Thought: A Journal of Philosophy*, **1**(4), 330–336.

Eichler, Michael. 2012. Causal inference in time series analysis. In: Berzuini, C., Dawid, P., and Bernardinelli, L., *Causality: Statistical perspectives and applications*. Hoboken, NJ: John Wiley & Sons, pp. 327–354.

Eichler, Michael, and Didelez, Vanessa. 2010. On Granger causality and the effect of interventions in time series. *Lifetime Data Analysis*, **16**(1), 3–32.

Elhai, Jon D., Levine, Jason C., Dvorak, Robert D., and Hall, Brian J. 2016. Fear of missing out, need for touch, anxiety and depression are related to problematic smartphone use. *Computers in Human Behavior*, **63**, 509–516.

Elwert, Felix, and Winship, Christopher. 2014. Endogenous selection bias: the problem of conditioning on a collider variable. *Annual Review of Sociology*, **40**, 31–53.

Engle, Robert F, and Granger, Clive W.J. 1987. Co-integration and error correction: representation, estimation, and testing. *Econometrica: Journal of the Econometric Society*, 251–276.

Evans, Peter W. 2017. Quantum causal models, faithfulness, and eetrocausality. *The British Journal for the Philosophy of Science*, **69**(3), 745–774.

Faye, Jan. 2015. Empiricism. In: Zalta, Edward N. (ed), *The Stanford Encyclopedia of Philosophy*. Metaphysics Research Lab, Stanford University. http://plato.stanford.edu/archives/win2014/entries/peirce/.

Fedorov, Valerii Vadimovich. 1972. *Theory of optimal experiments*. New York: Academic Press.

Feldbaum, A.A. 1960. Dual control theory I–IV. *Avtomatika i Telemekhanika*.

Fernbach, Philip M, and Sloman, Steven A. 2009. Causal learning with local computations. *Journal of Experimental Psychology: Learning, Memory & Cognition*, **35**(3), 678.

Field, Hartry. 2003. Causation in a physical world. In: Loux, Michael J., and Zimmerman, Dean W. (eds), *The Oxford Handbook of Metaphysics*. Oxford: Oxford University Press, pp. 435–460.

Fieser, James. 2018. David Hume (1711–1776). In: *The Internet Encyclopedia of Philosophy*. https://www.iep.utm.edu/hume/.

Fisher, Franklin M. 1970. A correspondence principle for simultaneous equation models. *Econometrica*, **38**(1), 73–92.

Flew, Antony. 1956. Effects before their causes?—addenda and corrigenda. *Analysis*, **16**(5), 104–110.

— 1957. Causal disorder again. *Analysis*, **17**(4), 81–86.

Forster, Malcolm, Raskutti, Garvesh, Stern, Reuben, and Weinberger, Naftali. 2017. The frugal inference of causal relations. *The British Journal for the Philosophy of Science*. **69**(3), 821–848.

Freedman, David A. 1997. From association to causation via regression. In: McKim, Vaughn R., and Turner, Stephen P. (eds), *Causality in Crisis? Statistical Methods and the Search for Causal Knowledge in the Social Sciences*. Notre Dame, IN: University of Notre Dame Press, 113–61.

— 2009. *Statistical Models: Theory and Practice*. Revised edn. Cambridge: Cambridge University Press.

Friedman, William J. 1993. Memory for the Time of Past Events. *Psychological Bulletin*, **113**(1), 44–66.

Frisch, Mathias. 2014. *Causal Reasoning in Physics*. Cambridge: Cambridge University Press.

Friston, Karl J., Bastos, André M., Oswal, Ashwini, van Wijk, Bernadette, Richter, Craig, and Litvak, Vladimir. 2014. Granger causality revisited. *NeuroImage*, **101**, 796–808.

Frosch, Caren A., McCormack, Teresa, Lagnado, David A, and Burns, Patrick. 2012. Are causal structure and intervention judgments inextricably linked? A developmental study. *Cognitive Science*, **36**, 261–285.

Futch, Michael J. 2008. *Leibniz's Metaphysics of Time and Space*. New York: Springer.

Galavotti, Maria Carla. 2018. Wesley Salmon. In: Zalta, Edward N. (ed), *The Stanford Encyclopedia of Philosophy*. https://plato.stanford.edu/entries/wesley-salmon/CausProc.

Gale, Richard M. 1965. Why a cause cannot be later than its effect. *The Review of Metaphysics*, **19**(2), 209–234.

Gallistel, C. R., and Gibbon, John. 2000. Time, rate, and conditioning. *Psychological Review*, **107**(2), 289.

Gandini, Sara, Botteri, Edoardo, Iodice, Simona, Boniol, Mathieu, Lowenfels, Albert B., Maisonneuve, Patrick, and Boyle, Peter. 2008. Tobacco smoking and cancer: a meta-analysis. *International Journal of Cancer*, **122**(1), 155–164.

Garant, Dan, and Jensen, David. 2016. Evaluating causal models by comparing interventional distributions. *CoRR*. http://arxiv.org/abs/1608.04698

Garrett, Brian. 2014. Black on backwards causation. *Thought: A Journal of Philosophy*, **3**(3), 230–233.

— 2015. On the epistemic bilking argument. *Thought: A Journal of Philosophy*, **4**(3), 139–140.

Gastil, Raymond D. 1978. *Freedom in the World: Political Rights and Civil Liberties*. Boston, MA: Freedom House.

Gerstenberg, T., Goodman, Noah D., Lagnado, David A., and Tenenbaum, Joshua B. 2015. How, whether, why: Causal judgments as counterfactual contrasts. In: *Proceedings of the 37th Annual Meeting of the Cognitive Science Society*. Austin, TX: Cognitive Science Society, pp. 782–787.

Gerstenberg, Tobias, and Tenenbaum, Joshua B. 2017. Intuitive theories. Pages 515–548 of: Waldmann, Michael R. (ed), *The Oxford Handbook of Causal Reasoning*. New York: Oxford University Press.

Gerstenberg, Tobias, Bechlivanidis, Christos, and Lagnado, David A. 2013. Back on track: backtracking in counterfactual reasoning. In: *Proceedings of the 35th Annual Meeting of the Cognitive Science Society*. Austin, TX: Cognitive Science Society.

Getoor, Lise, and Taskar, Ben. 2007. *Introduction to Statistical Relational Learning*. Cambridge, MA: The MIT press.

Getoor, Lise, Friedman, Nir, Koller, Daphne, and Taskar, Benjamin. 2001. Learning Probabilistic Models of Relational Structure. In: *Proceedings of the Eighteenth International Conference on Machine Learning (ICML-'01)*, pp. 170–177.

Ghahramani, Zoubin. 1998. Learning dynamic Bayesian networks. In: Giles, C. Lee, and Gori, Marco (eds), *Adaptive Processing of Sequences and Data Structures*. Berlin, Heidelberg: Springer Berlin Heidelberg, pp. 168–197.

Giere, Ronald. 1997. *Understanding Scientific Reasoning (4th edition)*. Fort Worth: Harcourt Brace College Publishers.

Gijsbers, Victor. On the presuppositions of (nearly) all current theories of causation. *Under Review*.

Glennan, Stuart. 1996a. Mechanisms and the nature of causation. *Erkenntnis*, **44**(1), 49–71.

— 1996b. Mechanisms and the Nature of Causation. *Erkenntnis*, **44**(1), 49–71.

— 2009. Productivity, relevance and natural selection. *Biology & Philosophy*, **24**(3), 325–339.

— 2017. *The New Mechanical Philosophy*. New York: Oxford University Press.

Glymour, Clark. 1999. Rabbit hunting. *Synthese*, **121**(1), 55–78.

— 2001. *The Mind's Arrows: Bayes Nets and Graphical Causal Models in Psychology*. Cambridge, MA: The MIT Press.

Glymour, Clark, Scheines, Richard, Spirtes, Peter, and Kelley, Kevin. 1987. *Discovering Causal Structure: Artificial Intelligence, Philosophy of Science, and Statistical Modeling*. Orlando: Academic Press, Inc.

Glymour, Clark, Scheines, Richard, Spirtes, Peter, and Meek, Christopher. 1994. "Regression and Causation". Tech. rept. CMU-PHIL-60. Pittsburgh, PA.

Glymour, Clark, and Eberhardt, Frederick. 2016. Hans Reichenbach. In: Zalta, Edward N. (ed), *The Stanford Encyclopedia of Philosophy*. https://plato.stanford.edu/entries/reichenbach/DirTim195.

Goodman, Noah, Mansinghka, Vikash, Roy, Daniel M., Bonawitz, Keith, and Tenenbaum, Joshua B. 2012. Church: a language for generative models. *arXiv preprint arXiv:1206.3255*.

Gopnik, Alison, Glymour, Clark, Sobel, David M., Schulz, Laura E., Kushnir, Tamar, and Danks, David. 2004. A theory of causal learning in children: causal maps and Bayes nets. *Psychological Review*, **111**(1), 3.

Gopnik, Alison, and Schulz, Laura E. 2007. *Causal learning: Psychology, philosophy, and computation*. New York: Oxford University Press.

Gopnik, Alison, and Sobel, David M. 2000. Detecting blickets: how young children use information about novel causal powers in categorization and induction. *Child Development*, **71**(5), 1205–1222.

Gordon, Andrew D., Henzinger, Thomas A., Nori, Aditya V., and Rajamani, Sriram K. 2014. Probabilistic programming. In: *Proceedings of the on Future of Software Engineering*. ACM, pp. 167–181.

Granger, Clive W.J. 1969. Investigating causal relations by econometric models and cross-spectral methods. *Econometrica*, **37**(3), 424–438.

— 1980. Testing for causality: a personal viewpoint. *Journal of Economic Dynamics and Control*, **2**, 329–352.

— 2004. Time series analysis, cointegration, and applications. *American Economic Review*, 421–425.

Green, Sara, and Batterman, Robert. 2017. Biology meets physics: reductionism and multi-scale modeling of morphogenesis. *Studies in History and Philosophy of Science Part C: Studies in History and Philosophy of Biological and Biomedical Sciences*, **61**, 20–34.

Greenwald, Anthony G. 1976. Within-subjects designs: to use or not to use? *Psychological Bulletin*, **83**(2), 314–320.

Greville, W. James, and Buehner, Marc J. 2007. The influence of temporal distributions on causal induction from tabular data. *Memory & Cognition*, **35**(3), 444–453.

— 2010. Temporal predictability facilitates causal learning. *Journal of Experimental Psychology: General*, **139**(4), 756–771.

— 2016. Temporal predictability enhances judgements of causality in elemental causal induction from both observation and intervention. *The Quarterly Journal of Experimental Psychology*, **69**(4), 678–697.

Greville, W. James, Cassar, Adam A., Johansen, Mark K., and Buehner, Marc J. 2013. Structural awareness mitigates the effect of delay in human causal learning. *Memory & Cognition*, **41**(6), 1–13.

Grice, G Robert. 1948. The relation of secondary reinforcement to delayed reward in visual discrimination learning. *Journal of Experimental Psychology*, **38**(1), 1.

Griffiths, Thomas L., Lucas, Christopher G., Williams, Joseph, and Kalish, Michael L. 2009. Modeling human function learning with Gaussian processes. In: *Advances in Neural Information Processing Systems*, pp. 553–560.

Griffiths, Thomas L. 2005. Causes, coincidences, and theories. Ph.D. thesis, Stanford University.

Griffiths, Thomas L., and Tenenbaum, Joshua B. 2005. Structure and strength in causal induction. *Cognitive Psychology*, **51**(4), 334–384.

— 2009. Theory-based causal induction. *Psychological Review*, **116**(4), 661–716.

Grünbaum, Adolf. 1963. *Philosophical Problems of Space and Time*. New York: Alfred A. Knopf.

— 1968. *Modern Science and Zeno's Paradoxes*. Middletown, CT: Wesleyan University Press.

Guez, Arthur. 2015. Sample-based Search Methods for Bayes-Adaptive Planning. Ph.D. thesis, University College London.

Haggard, Patrick, Clark, Sam, and Kalogeras, Jeri. 2002. Voluntary action and conscious awareness. *Nature Neuroscience*, **5**(4), 382–385.

Hagmayer, York, and Waldmann, Michael R. 2002. How temporal assumptions influence causal judgments. *Memory & Cognition*, **30**(7), 1128–1137.

Hahn, Jinyong, Todd, Petra, and Van der Klaauw, Wilbert. 2001. Identification and estimation of treatment effects with a regression-discontinuity design. *Econometrica*, **69**(1), 201–209.

Hall, Ned. 2004. Two Concepts of Causation. In: Collins, John, Hall, Ned, and Paul, Laurie (eds), *Causation and Counterfactuals*. Cambridge, MA: MIT Press, pp. 225–276.

Halpern, Joseph Y. 2016. *Actual Causality*. Cambridge, MA: MIT Press.

Hausman, Daniel M. 1998. *Causal Asymmetries*. Cambridge: Cambridge University Press.

Hausman, Daniel M., and Woodward, James. 1999. Independence, invariance and the causal Markov condition. *The British Journal for the Philosophy of Science*, **50**(4), 521–583.

Hill, Austin Bradford. 1965. The environment and disease: association or causation? *Proceedings of the Royal Society of Medicine*, **58**(5), 295–300.

Hinkle, Dennis E., Wiersma, William, and Jurs, Stephen G. 1988. *Applied Statistics for the Behavioral Sciences (2nd edition)*. Boston, MA: Houghton Mifflin Company.

Hitchcock, Christopher. 2001. The intransitivity of causation revealed in equations and graphs. *The Journal of Philosophy*, **98**(6), 273–299.

— 2012. Events and times: a case study in means–ends metaphysics. *Philosophical studies*, **160**(1), 79–96.

— 2016. Probabilistic causation. In: Zalta, Edward N. (ed), *The Stanford Encyclopedia of Philosophy*, winter 2016 edn. Metaphysics Research Lab, Stanford University.

Hoefer, Carl. 2009. Causation in spacetime theories. In: Beebee, Helen, Hitchcock, Christopher Read, and Menzies, Peter (eds), *The Oxford Handbook of Causation*. Oxford: Oxford University Press, pp. 687–706.

Holland, Paul W. 1986. Statistics and causal inference. *Journal of the American Statistical Association*, **81**(396), 945–960.

Holyoak, Keith J., and Cheng, Patricia W. 2011. Causal learning and inference as a rational process: the new synthesis. *Annual Review of Psychology*, **62**, 135–63.

Hoover, Kevin D. 2001. *Causality in Macroeconomics*. Cambridge: Cambridge University Press.

— 2003. Nonstationary time series, cointegration, and the principle of the common cause. *The British Journal for the Philosophy of Science*, **54**(4), 527–551.

— 2015. The ontological status of shocks and trends in macroeconomics. *Synthese*, **192**(11), 3509–3532.

Huemer, Michael, and Kovitz, Ben. 2003. Causation as simultaneous and continuous. *The Philosophical Quarterly*, **53**(213), 556–565.

Hume, David. 1739/1978. *A Treatise of Human Nature*. Oxford: Clarendon Press. Edited by L. A. Selby-Bigge & P. H. Nidditch.

— 1740/1978. *An Abstract of A Treatise of Human Nature*. Oxford: Clarendon Press. Edited by L. A. Selby-Bigge & P. H. Nidditch.

— 1748. *An Enquiry Concerning Human Understanding*. Ed. T. Beauchamp 2000. Oxford: Oxford University Press.

Ilhan, Zehra Esra, DiBaise, John K., Isern, Nancy G., Hoyt, David W., Marcus, Andrew K., Kang, Dae-Wook, Crowell, Michael D., Rittmann, Bruce E., and Krajmalnik-Brown, Rosa. 2017. Distinctive microbiomes and metabolites linked with weight loss after gastric bypass, but not gastric banding. *The ISME Journal*, **11**(9), 2047–2058.

Illari, Phyllis, and Russo, Federica. 2014. *Causality: Philosophical Theory Meets Scientific Practice*. Oxford: Oxford University Press.

Illari, Phyllis McKay. 2011. Mechanistic evidence: disambiguating the Russo-Williamson thesis. *International Studies in the Philosophy of Science*, **25**(2), 139–157.

Illari, Phyllis McKay, and Williamson, Jon. 2012. What is a mechanism? Thinking about mechanisms across the sciences. *European Journal for Philosophy of Science*, **2**(1), 119–135.

Imai, Kosuke, Keele, Luke, Tingley, Dustin, and Yamamoto, Teppei. 2011. Unpacking the black box of causality: learning about causal mechanisms from experimental and observational studies. *American Political Science Review*, **105**(4), 765–789.

Ismael, Jenann. 2016. How do causes depend on us? *Synthese*, **193**(1), 245–267.

Iwasaki, Yumi, and Simon, Herbert A. 1994. Causality and model abstraction. *Artificial intelligence*, **67**(1), 143–194.

James, William. 1890. *The Principles of Psychology*. Dover (2000 reprint).

Janiak, Andrew. 2013. Three concepts of causation in Newton. *Studies in History and Philosophy of Science*, **44**(3), 396–407.

Janiak, Andrew, and Schliesser, Eric (eds). 2012. *Interpreting Newton: Critical Essays*. Cambridge: Cambridge University Press.

Jensen, David D, and Cohen, Paul R. 2000. Multiple comparisons in induction algorithms. *Machine Learning*, **38**(3), 309–338.

Jordan, Michael I, and Rumelhart, David E. 1992. Forward models: supervised learning with a distal teacher. *Cognitive Science*, **16**(3), 307–354.

Jung, Carl. 1973. *Synchronicity: An Acausal Connecting Principle*. New Haven, CT: Princeton University Press.

Kant, Immanuel. 1781/1998. *Critique of Pure Reason*. Cambridge: Cambridge University Press.

Karin, Michael. 1999. How NF-κB is activated: the role of the IκB kinase (IKK) complex. *Oncogene*, **18**, 6867–6874.

Kim, Nancy S, and Ahn, Woo-Kyoung. 2002. Clinical psychologists' theory-based representations of mental disorders predict their diagnostic reasoning and memory. *Journal of Experimental Psychology: General*, **131**(4), 451.

Kirby, James B. 2002. The influence of parental separation on smoking initiation in adolescents. *Journal of Health and Social Behavior*, **43**(1), 56–71.

Klenske, Edgar D., and Hennig, Philipp. 2016. Dual control for approximate bayesian reinforcement learning. *Journal of Machine Learning Research*, **17**(127), 1–30.

Kline, Rex B. 2011. *Principles and Practice of Structural Equation Modeling*. 3rd edn. New York: The Guilford Press.

Koller, Daphne, and Friedman, Nir. 2009. *Probabilistic Graphical Models: Principles and Techniques*. Cambridge, MA: MIT press.

Korb, Kevin B., Hope, Lucas R., Nicholson, Ann E., and Axnick, Karl. 2004. Varieties of causal intervention. In: *Pacific Rim International Conference on Artificial Intelligence*. Berlin, Heidelberg: Springer, pp. 322–331.

Korman, Daniel Z. 2014. The Vagueness Argument against Abstract Artifacts. *Philosophical Studies*, **167**(1), 57–71.

Kreuz, Sebastian, Siegmund, Daniela, Rumpf, Jost-Julian, Samel, Dierk, Leverkus, Martin, Janssen, Ottmar, Häcker, Georg, Dittrich-Breiholz, Oliver, Kracht, Michael, Scheurich, Peter, and Wajant, Harald. 2004. NFκB activation by Fas is mediated through FADD, caspase-8, and RIP and is inhibited by FLIP. *The Journal of Cell Biology*, **166**(3), 369–380.

Kutach, Douglas. 2014. *Causation*. Cambridge, UK: Polity.

Lacey, Hugh M. 1968. The causal theory of time: a critique of Grünbaum's version. *Philosophy of Science*, **35**(4), 332–354.

Lacko, Vladimír. 2012. Planning of experiments for a nonautonomous Ornstein-Uhlenbeck process. *Tatra Mountains Mathematical Publications*, **51**(1), 101–113.

Lagnado, David A, and Sloman, Steven A. 2002. Learning causal structure. In: *Proceedings of the 24th Annual Meeting of the Cognitive Science Society*. Erlbaum.

— 2004. The advantage of timely intervention. *Journal of Experimental Psychology: Learning, Memory & Cognition*, **30**, 856–876.

— 2006. Time as a guide to cause. *Journal of Experimental Psychology: Learning, Memory, and Cognition*, **32**(3), 451–460.

Lagnado, David A, and Speekenbrink, Maarten. 2010. The influence of delays in real-time causal learning. *The Open Psychology Journal*, **3**(2), 184–195.

Lagnado, David A, Gerstenberg, Tobias, and Zultan, Ro'i. 2013. Causal responsibility and counterfactuals. *Cognitive Science*, **37**(6), 1036–1073.

Lake, Brenden M, Salakhutdinov, Ruslan, and Tenenbaum, Joshua B. 2015. Human-level concept learning through probabilistic program induction. *Science*, **350**(6266), 1332–1338.

Lauritzen, Steffen L, and Richardson, Thomas S. 2002. Chain graph models and their causal interpretations. *Journal of the Royal Statistical Society*, **64**(3), 321–348.

Lee, Sanghack, and Honavar, Vasant. 2016a. A characterization of Markov equivalence classes of relational causal models under path semantics. In: *Proceedings of the Thirty-Second Conference on Uncertainty in Artificial Intelligence (UAI-16)*.

— 2016b. On learning causal models from relational data. In: *Proceedings of the Thirtieth AAAI Conference on Artificial Intelligence (AAAI-16)*, pp. 3263–3270.

Leuridan, Bert. 2012. Three problems for the mutual manipulability account of constitutive relevance in mechanisms. *The British Journal for the Philosophy of Science*, **63**(2), 399–427.

Lewis, David. 1972. Psychophysical and theoretical identifications. *Australasian Journal of Philosophy*, **50**(3), 249–258.

— 1973a. Causation. *The Journal of Philosophy*, **70**(17), 556–567.

— 1980. A subjectivist's guide to objective chance. In: Jeffrey, Richard C. (ed), *Studies in Inductive Logic and Probability, Volume II*. Berkeley: University of California Press, pp. 263–293.

— 1979. Counterfactual dependence and time's arrow. *Noûs*, **13**(4), 455–476.

— 1986. A subjectivist's guide to objective chance. In: *Philosophical Papers*, vol. 2. Oxford: Oxford University Press, pp. 83–132.

— 1973b. *Counterfactuals*. Cambridge, MA: Harvard University Press.

Little, Daniel. 2012. Explanatory autonomy and Coleman's boat. *THEORIA. Revista de Teoría, Historia y Fundamentos de la Ciencia*, **27**(2), 137–151.

Lucas, Christopher G, and Kemp, Charles. 2015. An improved probabilistic account of counterfactual reasoning. *Psychological Review*, **122**(4), 700.

McCormack, Teresa, Bramley, Neil R., Frosch, Caren A., Patrick, Fiona, and Lagnado, David A. 2016. Children's use of interventions to learn causal structure. *Journal of Experimental Child Psychology*, **141**, 1–22.

Machamer, Peter, Darden, Lindley, and Craver, Carl F. 2000. Thinking about mechanisms. *Philosophy of Science*, **67**(1), 1–25.

Mackie, John L. 1974/1980. *The Cement of the Universe*. Oxford: Clarendon Press.

Maier, Marc. 2014.   Causal Discovery for Relational Domains: Representation, Reasoning, and Learning. Ph.D. thesis, University of Massachusetts Amherst.

Maier, Marc, Marazopoulou, Katerina, Arbour, David, and Jensen, David. 2013. A sound and complete algorithm for learning causal models from relational data. In: *Proceedings of the Twenty-Ninth Conference on Uncertainty in Artificial Intelligence*. AUAI Press, pp. 371–380.

Malthus, Thomas Robert. 1888. *An Essay on the Principle of Population: or, A View of its Past and Present Effects on Human Happiness*. London: Reeves & Turner.

Mandler, George, and Shebo, Billie J. 1982. Subitizing: an analysis of its component processes. *Journal of Experimental Psychology: General*, **111**(1), 1.

Mansinghka, Vikash, Selsam, Daniel, and Perov, Yura. 2014. Venture: a higher-order probabilistic programming platform with programmable inference. *arXiv preprint arXiv:1404.0099*.

Marazopoulou, Katerina, Maier, Marc, and Jensen, David. 2015. Learning the structure of causal models with relational and temporal dependence. In: *Proceedings of the Thirty-First Conference on Uncertainty in Artificial Intelligence (UAI-15)*, pp. 66–75.

Marom, Shimon. 2010. Neural timescales or lack thereof. *Progress in Neurobiology*, **90**(1), 16–28.

Maudlin, Tim. 2007. *The Metaphysics within Physics*. Oxford: Oxford University Press.

Mayr, Ernst. 1961. Cause and effect in biology. *Science*, **134**(3489), 1501–1506.

Mayrhofer, Ralf, and Waldmann, Michael R. 2016. Sufficiency and necessity assumptions in causal structure induction. *Cognitive Science*, **40**(8), 2137–2150.

Mehlberg, Henry. 1980. *Time, Causality and the Quantum Theory. Volume 1: Essay on the Causal Theory of Time*. Dordrecht: Springer.

Mellor, D. H. 1991. Causation and the direction of time. *Erkenntnis*, **35**(1), 191–203.

Menzies, Peter. 1996. Probabilistic causation and the pre-emption problem. *Mind*, **105**(417), 85–117.

— 2014.   Counterfactual theories of causation.   In: Zalta, Edward N. (ed), *The Stanford Encyclopedia of Philosophy*, spring 2014 edn. Metaphysics Research Lab, Stanford University.

Menzies, Peter, and Price, Huw. 1993. Causation as a secondary quality. *The British Journal for the Philosophy of Science*, **44**(2), 187–203.

Michotte, Albert. 1946. *La Perception de la Causalité*. Louvain: Editions de l'Institut Supérieur de Philosophie.

Miele, Christopher. 2016. Pushing that crosswalk button may make you feel better, but . . . . *New York Times*, October 27.

Mill, John S. 1843/1911. *A System of Logic: Ratiocinative and Inductive*. London: Longmans, Green & Co.

Miller, Warren E. 1999. Temporal order and causal inference. *Political Analysis*, **8**(2), 119–146.

Mitchell, Sandra D. 2003.   *Biological Complexity and Integrative Pluralism*. Cambridge: Cambridge University Press.

Momennejad, Ida, Russek, Evan M., Cheong, Jin H., Botvinick, Matthew M., Daw, N. D., and Gershman, Samuel J. 2017. The successor representation in human reinforcement learning. *Nature Human Behaviour*, **1**(9), 680–692.

Morris, William Edward, and Brown, Charlotte R. 2017. David Hume. In: Zalta, Edward N. (ed), *The Stanford Encyclopedia of Philosophy*, spring 2017 edn. Metaphysics Research Lab, Stanford University.

Mumford, Stephen, and Anjum, Rani Lill. 2011. *Getting Causes from Powers*. Oxford: Oxford University Press.

Murphy, Kevin. 2002. Dynamic Bayesian Networks: Representation, Inference and Learning. Ph.D. thesis, University of California, Berkley.

Nikolic, Milena, and Lagnado, David A. 2015. There aren't plenty more fish in the sea: a causal network approach. *British Journal of Psychology*, **106**(4), 564–582.

Nodelman, Uri, Shelton, Christian R., and Koller, Daphne. 2002. Continuous time Bayesian networks. In: *Proceedings of the 18th Conference on Uncertainty in Artificial Intelligence*. San Francisco, CA: Morgan Kaufmann, pp. 378–387.

North, Jill. 2008. Two views on time reversal. *Philosophy of Science*, **75**(2), 201–223.

Oates, Tim, and Jensen, David. 1999. Toward a theoretical understanding of why and when decision tree pruning algorithms fail. In: *Proceedings of the sixteenth national conference on Artificial intelligence and the eleventh Innovative applications of artificial intelligence conference innovative applications of artificial intelligence*, pp. 372–378.

Odum, Eugene P. 1959. *Fundamentals of Ecology*. Philadelphia: W.B. Saunders company.

Pacer, Michael D, and Griffiths, Thomas L. 2011. A rational model of causal induction with continuous causes. In: *Advances in Neural Information Processing Systems*, pp. 2384–2392.

— 2015. Upsetting the contingency table: causal induction over sequences of point events. In: *Proceedings of the 37th Annual Meeting of the Cognitive Science Society*. Austin, TX: Cognitive Science Society.

Paek, Andrew L., Liu, Julia C., Loewer, Alexander, Forrester, William C., and Lahav, Galit. 2016. Cell-to-Cell Variation in p53 Dynamics Leads to Fractional Killing. *Cell*, **165**(3), 631–642.

Park, Gunwoong, and Raskutti, Garvesh. 2016. Identifiability assumptions and algorithm for directed graphical models with feedback. *arXiv preprint arXiv:1602.04418*.

Paudel, Jayash, and Crago, Christine L. 2017. *Fertilizer Subsidy and Agricultural Productivity: Empirical Evidence from Nepal*. Tech. rept. Agricultural and Applied Economics Association.

Paul, L. A., and Hall, Ned. 2013. *Cases that Threaten Transitivity*. Oxford: Oxford University Press.

Pearl, Judea. 1988. *Probabilistic Reasoning in Intelligent Systems*. San Francisco, CA: Morgan Kaufmann.

— 2000. *Causality: Models, Reasoning, and Inference*. Cambridge: Cambridge University Press.

— 2009. Causal inference in statistics: an overview. *Statistics Surveys*, **3**, 96–146.

— 2014. *Probabilistic Reasoning in Intelligent Systems: Networks of Plausible Inference*. San Francisco: Morgan Kaufmann.

Pfeffer, Avi. 2016. *Practical Probabilistic Programming*. Shelter Island, NY: Manning Publications Co.

Pfeiffer, Brad E, and Foster, David J. 2013. Hippocampal place-cell sequences depict future paths to remembered goals. *Nature*, **497**(7447), 74–79.

Piaget, Jean, and Valsiner, Jaan. 1930. *The Child's Conception of Physical Causality*. London: Routledge

Price, Huw. 1991. Agency and probabilistic causality. *The British Journal for the Philosophy of Science*, **42**(2), 157–176.

— 1992. Agency and causal asymmetry. *Mind*, **101**(403), 501–520.

— 1996. *Time's Arrow & Archimedes' Point*. Oxford: Oxford University Press.

— 2005. Causal Perspectivalism. In: Price, Huw, and Corry, Richard (eds), *Causation, Physics, and the Constitution of Reality: Russell's Republic Revisited*. Oxford: Oxford University Press, pp. 250–292.

— 2011. *Naturalism Without Mirrors*. New York: Oxford University Press.

— 2012a. Causation, chance, and the rational significance of supernatural evidence. *Philosophical Review*, **121**(4), 483–538.

— 2012b. Does time-symmetry imply retrocausality? How the quantum world says "maybe". *Studies in History and Philosophy of Science Part B: Studies in History and Philosophy of Modern Physics*, **43**(2), 75–83.

— 2017. Causation, intervention and agency—Woodward on Menzies and Price. In: Beebee, Helen, Hitchcock, Christopher, and Menzies, Peter (eds), *Making a Difference: Essays on the Philosophy of Causation*. New York: Oxford University Press.

Psillos, Stathis. 2002. *Causation and Explanation*. Chesham: Acumen Publishing.

Raiffa, Howard. 1974. *Applied Statistical Decision Theory*. Boston, MA: Harvard University Press.

Rehder, Robert E. 2003. A causal-model theory of conceptual representation and categorization. *Journal of Experimental Psychology: Learning, Memory & Cognition*, **29**(6), 1141.

— 2017. Reasoning with causal cycles. *Cognitive Science*, **41**.

Rehder, Robert E., and Burnett, Russell C. 2005. Feature inference and the causal structure of categories. *Cognitive Psychology*, **50**(3), 264–314.

Rehder, Robert E., and Waldmann, Michael R. 2017. Failures of explaining away and screening off in described versus experienced causal learning scenarios. *Memory & Cognition*, **45**(2), 245–260.

Reichenbach, Hans. 1928. *Philosophie der Raum-Zeit-Lehre*. Walter de Gruyter.

— 1956. *The Direction of Time*. Berkeley: University of California Press. Reprint, Dover Publications, 2000.

— 1956/1971. *The Direction of Time*. Berkeley: University of California Press.

Reiss, Ira L. 1967. *The Social Context of Premarital Sexual Permissiveness*. New York: Holt, Rinehart, & Winston.

Reiss, Julian. 2007. Time series, nonsense correlations and the principle of the common cause. In Russo, Federica and Williamson, Jon, (eds.), *Causality and*

Probability in the Sciences. Texts in Philosophy (5). London: College Publications, pp. 179–196.

Rescorla, Robert A, and Wagner, Allan R. 1972. A theory of Pavlovian conditioning: variations in the effectiveness of reinforcement and nonreinforcement. In: Black, Abraham H, and Prokasy, William Frederick (eds), *Classical Conditioning II: Current Research and Theory.* New York: Appleton-Century-Crofts, pp. 64–99.

Richardson, Thomas. 1996. A discovery algorithm for directed cyclic graphs. In: *Proceedings of the Twelfth International Conference on Uncertainty in Artificial Intelligence.* San Francisco, CA: Morgan Kaufmann Publishers Inc., pp. 454–461.

Rips, Lance J. 2010. Two causal theories of counterfactual conditionals. *Cognitive Science,* **34**(2), 175–221.

Rips, Lance J., and Edwards, Brian J. 2013. Inference and explanation in counterfactual reasoning. *Cognitive Science,* **37**(6), 1107–1135.

Roache, Rebecca. 2009. Bilking the Bilking Argument. *Analysis,* **69**(4), 605–611.

Romero, Felipe. 2015. Why there isn't inter-level causation in mechanisms. *Synthese,* **192**(11), 3731–3755.

Rosen, Gideon. 2014. Abstract Objects. In: Zalta, Edward N. (ed), *The Stanford Encyclopedia of Philosophy,* fall 2014 edn.

Rosenbaum, Paul R, and Rubin, Donald B. 1983. The central role of the propensity score in observational studies for causal effects. *Biometrika,* **70**(1), 41–55.

Ross, Lauren N. 2018. Causal selection and the pathway concept. *Philosophy of Science,* **85**(4), 551–572.

Rothe, Anselm, Deverett, Ben, and Kemp, Charles. Successful structure learning from observational data. *Cognition,* **179,** 266–297.

Rottman, Benjamin M. 2016. Searching for the best cause: roles of mechanism beliefs, autocorrelation, and exploitation. *Journal of Experimental Psychology: Learning, Memory, and Cognition,* **42**(8), 1233–1256.

Rottman, Benjamin M., and Keil, Frank C. 2012. Causal structure learning over time: observations and interventions. *Cognitive Psychology,* **64**(1), 93–125.

Rottman, Benjamin M., Kominsky, Jonathan F., and Keil, Frank C. 2014. Children use temporal cues to learn causal directionality. *Cognitive Science,* **38**(3), 489–513.

Rubin, Donald B. 2005. Causal inference using potential outcomes: design, modeling, decisions. *Journal of the American Statistical Association,* **100**(469), 322–331.

Russell, Bertrand. 1913. On the notion of cause. *Proceedings of the Aristotelian Society,* **13**(1), 1–26.

— 1927. *The Analysis of Matter.* London: Kegan Paul.

Salmon, Wesley C. 1984. *Scientific Explanation and the Causal Structure of the World.* Princeton: Princeton University Press.

— 1990. Scientific explanation: causation and unification. *Critica,* **22**(66), 3–23.

— 1994. Causality without counterfactuals. *Philosophy of Science,* **61**(2), 297–312.

— 1997. Causality and explanation: a reply to two critiques. *Philosophy of Science,* **64**(3), 461–477.

Saraiva, Nuno, Prole, David L., Carrara, Guia, Johnson, Benjamin F., Taylor, Colin W., Parsons, Maddy, and Smith, Geoffrey L. 2013. hGAAP promotes cell adhesion and migration via the stimulation of store-operated Ca^{2+} entry and calpain 2. *The Journal of Cell Biology,* **202**(4), 699–713.

Schlottmann, Anne. 1999. Seeing it happen and knowing how it works: how children understand the relation between perceptual causality and underlying mechanism. *Developmental Psychology*, **35**(5), 303–317.

Schulz, Eric, Klenske, Edgar D, Bramley, Neil R, and Speekenbrink, Maarten. 2017b. Strategic exploration in human adaptive control. In: *Proceedings of the 39th Annual Meeting of the Cognitive Science Society*. Austin, TX: Cognitive Science Society.

Schulz, Eric, Tenenbaum, Joshua B, Duvenaud, David, Speekenbrink, Maarten, and Gershman, Samuel J. 2017a. Compositional inductive biases in function learning. *Cognitive Psychology*, **99**, 44–79.

Schulz, Laura E. 2001. Do-calculus? Adults and preschoolers infer causal structure from patterns of outcomes following interventions. In: *Second Biennial Meeting of the Cognitive Development Society*.

Scriven, Michael. 1956. Randomness and the causal order. *Analysis*, **17**(1), 5–9.

Shadish, William, Cook, Thomas D., and Campbell, Donald T. 2001. *Experimental and Quasi-Experimental Designs for Generalized Causal Inference*. Boston: Houghton Mifflin.

Shanks, David R, and Dickinson, Anthony. 1987. Associative accounts of causality judgment. *The Psychology of Learning and Motivation*, **21**, 229–261.

Shanks, David R., Pearson, Susan M., and Dickinson, Anthony. 1989. Temporal contiguity and the judgement of causality by human subjects. *The Quarterly Journal of Experimental Psychology*, **41 B**(2), 139–159.

Shannon, Claude E. 1951. Prediction and entropy of printed English. *The Bell System Technical Journal*, **30**, 50–64.

Simon, Herbert. 1977. Causal ordering and identifiability. In: Hood, W., and Koopmans, T. (eds), *Studies in Econometric Method*. New York: John Wiley & Sons, pp. 49–74.

Simons, Leslie Gordon, Burt, Callie Harbin, and Peterson, F. Ryan. 2009. The effect of religion on risky sexual behavior among college students. *Deviant Behavior*, **30**(5), 467–485.

Sloman, Steven A. 2005. *Causal Models: How People Think about the World and its Alternatives*. Oxford: Oxford University Press.

Sloman, Steven A., and Lagnado, David A. 2005. Do we "do"? *Cognitive Science*, **29**(1), 5–39.

Sloman, Steven A., Love, Bradley C., and Ahn, Woo-Kyoung. 1998. Feature centrality and conceptual coherence. *Cognitive Science*, **22**(2), 189–228.

Smart, John J. C. 1969. Causal theories of time. *The Monist*, **53**(3), 385–395.

Smeenk, Chris, and Wüthrich, Christian. 2011. Time travel and time machines. In: Callender, Craig (ed), *The Oxford Handbook of Philosophy of Time*, online edn. Oxford: Oxford University Press.

Smith, Nicholas J. J. 2018. Time Travel. In: Zalta, Edward N. (ed), *The Stanford Encyclopedia of Philosophy*, summer 2018 edn. Metaphysics Research Lab, Stanford University.

Sobel, David M, and Kushnir, Tamar. 2006. The importance of decision making in causal learning from interventions. *Memory & Cognition*, **34**(2), 411–419.

Sobel, David M, Tenenbaum, Joshua B, and Gopnik, Alison. 2004. Children's causal inferences from indirect evidence: backwards blocking and Bayesian reasoning in preschoolers. *Cognitive Science*, **28**(3), 303–333.

Sobel, Michael E. 1990. Effect analysis and causation in linear structural equation models. *Psychometrika*, **55**(3), 495–515.

Sober, Elliott. 2001. Venetian sea levels, British bread prices, and the principle of the common cause. *The British Journal for the Philosophy of Science*, **52**(2), 331–346.

Soo, Kevin W, and Rottman, Benjamin M. 2014. Learning causal direction from transitions with continuous and noisy variables. In: *Proceedings of the Annual Meeting of the Cognitive Science Society*, vol. 36.

Soo, Kevin W, and Rottman, Benjamin M. 2018. Causal strength induction from time series data. *Journal of Experimental Psychology: General*, **147**(4), 485.

Spirtes, Peter, and Scheines, Richard. 1997. Reply to Freedman. In: McKim, Vaughn R., and Turner, Stephen P. (eds), *Causality in Crisis? Statistical Methods and the Search for Causal Knowledge in the Social Sciences*. Notre Dame, IN: University of Notre Dame Press, pp. 163–182.

Spirtes, Peter, Glymour, Clarke, and Scheines, Richard. 2000. *Causation, Prediction, and Search*. Second edn. Cambridge, MA: MIT Press. First published 1993.

Steel, Daniel. 2004. Social mechanisms and causal inference. *Philosophy of the Social Sciences*, **34**(1), 55–78.

— 2008. *Across the Boundaries: Extrapolation in Biology and Social Science*. New York: Oxford University Press.

Stephan, Simon, Mayrhofer, Ralf, and Waldmann, Michael R. 2018. Assessing singular causation: the role of causal latencies. In: *Proceedings of the 40th Annual Conference of the Cognitive Science Society*. Cognitive Science Society.

Steyvers, Mark, Lee, Michael D, and Wagenmakers, Eric-Jan. 2009. A Bayesian analysis of human decision-making on bandit problems. *Journal of Mathematical Psychology*, **53**(3), 168–179.

Steyvers, Mark, Tenenbaum, Joshua B, Wagenmakers, Eric-Jan, and Blum, Ben. 2003. Inferring causal networks from observations and interventions. *Cognitive Science*, **27**(3), 453–489.

Strawson, Peter. 1966. *The Bounds of Sense*. London: Methuen.

Strotz, Robert H., and Wold, H. O. A. 1960. Recursive vs. nonrecursive systems: an attempt at synthesis. *Econometrica*, **28**(2), 417–427.

Suppes, Patrick. 1970. *A Probabilistic Theory of Causality*. Amsterdam: North-Holland.

Swoyer, Chris. 1991. Structural representation and surrogative reasoning. *Synthese*, **87**(3), 449–508.

Taylor, Charles Lewis, and Jodice, David A. 1983. *World Handbook of Political and Social Indicators III. Volume 2: Political Protest and Government Change*. New Haven and London: Yale University Press.

Timberlake, Michael, and Williams, Kirk R. 1984. Dependence, political exclusion, and government repression: some cross-national evidence. *American Sociological Review*, **49**(1), 141–146.

Tolpin, David, van de Meent, Jan-Willem, and Wood, Frank. 2015. Probabilistic programming in Anglican. In: *Joint European Conference on Machine Learning and Knowledge Discovery in Databases*, pp. 308–311.

Tooley, Michael. 1997. *Time, Tense and Causation*. Oxford: Clarendon Press.

Tsamardinos, Ioannis, Brown, Laura E, and Aliferis, Constantin F. 2006. The max-min hill-climbing Bayesian network structure learning algorithm. *Machine Learning*, **65**(1), 31–78.

Tulving, Endel. 1972. Episodic and semantic memory. In: Tulving, Endel, and Donaldson, Wayne (eds), *Organization of memory*. Oxford: Academic Press, pp. 381–403.

Uffink, Jos. 2017. Boltzmann's work in statistical physics. In: Zalta, Edward N. (ed), *The Stanford Encyclopedia of Philosophy*, spring 2017 edn. Metaphysics Research Lab, Stanford University.

Uhlenbeck, George E, and Ornstein, Leonard S. 1930. On the theory of the Brownian motion. *Physical Review*, **36**(5), 823.

van de Meent, Jan-Willem, Paige, Brooks, Yang, Hongseok, and Wood, Frank. 2018. An introduction to probabilistic programming. *arXiv preprint arXiv:1809.10756*.

van Fraassen, Bas. 1970. *An Introduction to the Philosophy of Time and Space*. New York: Random House.

— 1972. Earman on the causal theory of time. *Synthese*, **24**(1–2): 87–95.

VanderWeele, Tyler J. 2009. Mediation and mechanism. *European Journal of Epidemiology*, **24**(5), 217–224.

Vernon, David and Lowe, Robert and Thill, Serge and Ziemke, Tom. 2015. Embodied cognition and circular causality: on the role of constitutive autonomy in the reciprocal coupling of perception and action. *Frontiers in Psychology*, **6**, 1660.

Verschuur, Gerrit L. 1996. *Hidden Attraction: The History and Mystery of Magnetism*. New York: Oxford University Press.

von Wright, Georg Henrik. 1971. *Explanation and Understanding*. New York: Cornell University Press.

Vul, Ed, Alvarez, George, Tenenbaum, Joshua B., and Black, Michael J. 2009. Explaining human multiple object tracking as resource-constrained approximate inference in a dynamic probabilistic model. In: *Advances in neural information processing systems*, pp. 1955–1963.

Waldmann, Michael R. 2000. Competition among causes but not effects in predictive and diagnostic learning. *Journal of Experimental Psychology: Learning, Memory & Cognition*, **26**(1), 53–76.

Waldmann, Michael R., and Holyoak, Keith J. 1992. Predictive and diagnostic learning within causal models: asymmetries in cue competition. *Journal of Experimental Psychology: General*, **121**(2), 222–236.

Waldmann, Michael R, and Martignon, Laura. 1998. A Bayesian network model of causal learning. In: *Proceedings of the twentieth annual conference of the Cognitive Science Society*, pp. 1102–1107.

Waldmann, Michael R., and Hagmayer, York. 2005. Seeing versus doing: two modes of accessing causal knowledge. *Journal of Experimental Psychology: Learning Memory & Cognition*, **31**(2), 216–227.

Wallace, David. 2012. *The Emergent Multiverse: Quantum Theory According to the Everett Interpretation*. Oxford: Oxford University Press.

Wheeler, John Archibald, and Feynman, Richard Phillips. 1949. Classical electrodynamics in terms of direct interparticle action. *Reviews of Modern Physics*, **21**(3), 425.

White, Peter A. 2006. How well is causal structure inferred from cooccurrence information? *European Journal of Cognitive Psychology*, **18**(3), 454–480.

— 2008. Beliefs about interactions between factors in the natural environment: a causal network study. *Applied Cognitive Psychology*, **22**(4), 559–572.

Wiseman, Howard M. 2014. The two Bell?s Theorems of John Bell. *Journal of Physics A: Mathematical and Theoretical*, **47**(42), 424001.

Wolfe, John B. 1921. The effect of delayed reward upon learning in the white rat. *Journal of Comparative Psychology*, **17**(1), 1–21.

Woodward, James. 2003. *Making Things Happen: A Theory of Causal Explanation*. Oxford: Oxford University Press.

— 2009. Agency and Interventionist Theories. In: Beebee, Helen, Hitchcock, Christopher, and Menzies, Peter (eds), *The Oxford Handbook of Causation*. Oxford: Oxford University Press, pp. 234–262.

— 2016. Causation and manipulability. In: Zalta, Edward N. (ed), *The Stanford Encyclopedia of Philosophy*. https://plato.stanford.edu/entries/causation-mani/AgenTheo.

— 2017. Interventionism and the missing metaphysics. In: Slater, Matthew, and Yudell, Zanja (eds), *Metaphysics and the Philosophy of Science: New Essays*. Oxford: Oxford University Press, pp. 193–228.

Wooldridge, Jeffrey M. 2015. *Introductory Econometrics: A Modern Approach*. Boston, MA : Cengage Learning Nelson Education.

Young, Michael E, and Nguyen, Nam. 2009. The problem of delayed causation in a video game: constant, varied, and filled delays. *Learning and Motivation*, **40**(3), 298–312.

Zhang, Jiji. 2012. A comparison of three Occam's razors for Markovian causal models. *The British Journal for the Philosophy of Science*, **64**(2), 423–448.

Zwart, P. J. 1967. *Causaliteit*. Assen: Van Gorcum.

Index